Organic Laboratory Techniques 3rd edition

Ralph J. Fessenden
University of Montana

■

(The Late) Joan S. Fessenden
University of Montana

■

Patty Feist
University of Colorado—Boulder

This text is accompanied by **WebWorks**, an online resource center dedicated to this text. For access to this helpful and practical resource, point your web browser to **http://www.brookscole.com** and search for your text under Resources. Follow the online instructions to create your own user name and password. You will be asked to supply the text ISBN number at the initial registration page. Your text ISBN number is **0-534-37981-8**.

Australia • Canada • Mexico • Singapore • Spain • United Kingdom • United States

BROOKS/COLE
THOMSON LEARNING™

Sponsoring Editor: *Jennifer Huber*
Marketing Team: *Carla Martin-Falcone, Stephanie Rogulski*
Editorial Assistant: *Brandon Horn*
Production Editor: *Stephanie Andersen*
Manuscript Editor: *Kristen Cassereau*
Permissions Editor: *Sue Ewing*
Cover Design: *Denise Davidson*
Cover Photo: *Anton Vengo/SuperStock*
Cover Coordinator: *Roy Neuhaus*
Print Buyer: *Kristine Waller*
Typesetting: *Patty Feist*
Printing and Binding: *Webcom Limited*

For more information about this or any other Brooks/Cole products, contact:
BROOKS/COLE
511 Forest Lodge Road
Pacific Grove, CA 93950 USA
www.brookscole.com
1-800-423-0563 (Thomson Learning Academic Resource Center)

Printed in Canada

10 9 8 7 6 5

Library of Congress Cataloging-in-Publication Data

Fessenden, Ralph J., 1932-
Organic laboratory techniques / Ralph J. Fessenden, Joan S. Fessenden, Patty Feist.—
3rd ed.
p. cm.
Includes index.
ISBN 0 534-37981-8 (paperback)
1. Chemistry, Organic—Laboratory manuals. I. Fessenden, Joan S. II. Feist, Patty, 1949- III. Title

QD261 .F466 2000
547'.0078—dc21 00-033718

Preface

Organic Laboratory Techniques, third edition, is a supplemental text for the introductory organic laboratory course in which the experiments are supplied by the instructor or in which the students work independently. In this text, the standard laboratory techniques are described. Since the techniques presented here are used in laboratory work by students of biochemistry, molecular biology, and cell biology, this text could be used in selected life science laboratory courses as well. It is also our hope that *Organic Laboratory Techniques*, Third Edition, finds its way into the professional library of every science student.

The general order of presentation is isolation and purification techniques, refractive index and chromatography, setting up and carrying out reactions, infrared and NMR spectroscopy, and use of the chemical literature. The order in which the techniques are used is flexible. For example, a student could use Technique 13 (gas chromatography) along with distillation (Techniques 5-8).

Each technique includes a brief theoretical discussion. Because many students in this course have not yet had a course in physical chemistry, we have kept these discussions fairly general. The steps used in each technique, along with common problems that might arise, are discussed in detail. Supplemental or related procedures are included for many of the techniques. Each chapter ends with a set of study problems that primarily stress the practical aspects of the technique. The reader is guided to additional problems and resources on the Brooks/Cole Web site.

We have emphasized safety throughout all the techniques. Safety notes are included with each technique where appropriate. Appendix III ("Health Hazards of Compounds used in Organic Chemistry") is included to alert the student to the possible toxicity of even common compounds.

This third edition has expanded, revised, or added information on the following topics.

Safety: The section on safety in the introduction to the text has been revised slightly to encourage students to both know their rights (per government protection agency regulations) and to take an active role in their own safety in the laboratory. The information on locating chemical hazard information has been moved to Appendix III.

Disposal of chemicals: Hazardous waste disposal is regulated by federal and state agencies, as interpreted and enforced by the safety departments of individual universities (and companies). Rather than give students exact instructions on waste disposal, we advise them to check with their instructor as to the proper methods to dispose of chemical wastes. References to printed and electronic sources of information on waste disposal policies and federal regulations are included on the Brooks/Cole Web site.

Microscale: The separate chapter on microscale has been removed. Instead, microscale techniques have been included in each technique chapter as appropriate. Our discussions of microscale techniques are not intended to be exhaustive. Proper coverage would require a separate technique book for each different style of microscale glassware. Most teaching facilities are scaling down if not to actual microscale, at least to smaller-scale reactions, in accordance with today's awareness of the impact of hazardous materials on the environment and on personal safety. Often conversion to microscale experiments can be a natural progression to the use of smaller-sized glassware rather than a drastic change in a university's laboratory experiments. Our intent is to provide the instructor with a nucleus of information upon which to build a smaller-scale laboratory program.

Chromatographic techniques: The chromatography chapters have been reorganized. Column chromatography is now covered first, with an expanded introduction to the general concepts of adsorbents and eluents pertinent to column, thin-layer, and gas chromatography. The flash column chromatography section has been expanded to include a flash chromatography method that can be performed without special equipment, and thus is readily suitable to undergraduate teaching laboratories. HPLC is only briefly covered because in reality, very few undergraduate organic chemistry teaching labs have access to this instrumentation. However, the HPLC section written for the second edition is available as a downloadable file on the Brooks/Cole Web site. The gas chromatography chapter has been revised to reflect the types of recorders currently used in undergraduate teaching laboratories.

Spectroscopy: The IR and NMR sections not included in the second edition have been returned to the third edition because of high demand. These sections have been changed to reflect the modernization of instrumentation to FT-IRs and FT-NMRs. Preparation of samples for IR spectroscopy has been expanded to include a simple sampling method for organic solids. Spectroscopy sections deleted from the second edition (the use of shift reagents, ringing and side bands in NMR, UV/vis spectroscopy) are available as downloadable files on the Brooks/Cole Web site.

Chemical literature: Much of the information (such as physical data) found in traditional, printed handbooks can now be found electronically. While the purist might argue that the most reliable source of information is contained in peer-reviewed printed tomes, a student is much more likely to search for this information from his or her own personal computer via a user-friendly interface than to travel to the academic library. Also, the Internet will be a ready tool for the student when they are no longer tied to an academic institution and the libraries therein. Therefore, the chemical literature chapter (Technique 17) lists printed sources *and* directs the student to useful Web sites.

The primary sources of chemical literature—research journals—have not changed. However, the method of retrieving this information from the abstract journals *is* rapidly changing. While print versions of the abstract journals are still available, researchers are increasingly taking advantage of electronic versions of abstract journals. Little special training is necessary to use these online databases because they are searchable in intuitive ways with which students are already familiar. Thus, we give the student a general overview of the types of information that are available and then direct them to their academic library for details on current offerings.

Toxicology: The appendix coverage on toxicology has been renamed "Health Hazards of Compounds Used in Organic Chemistry." This appendix now covers not only *where* to find chemical hazard information (MSDS's, bottle labels, and reference books or Web sites), but the *meaning* of hazard warnings. As in the chemical literature chapter, students are directed to both printed and electronic sources of information. The appendix also discusses the student's right to know what chemicals are stored in the facility and their right to know the hazards of these chemicals.

Miscellaneous changes: The lists of suggested readings have been moved to the Brooks/Cole Web site. These suggested readings have been expanded to include both additional printed sources and links to pertinent Internet sites. Last, but not least, we hope the numerous small changes made throughout the text have improved its readability.

We are indebted to James Hagen, University of Nebraska, Omaha; Todd Griffin, University of Indiana, Bloomington; and Raymond Lutz, Portland State University who have provided extensive and extraordinarily useful comments. Credit for some of the glassware illustrations new to the third edition goes to ChemClipArt 2000 by Molecular Arts Corporation (Anaheim, California).

Ralph J. Fessenden
Patty L. Feist

I would like to thank Tad Koch, Professor of Chemistry at the University of Colorado, Boulder, for his encouragement to take on this project in the first place and his input during the revision process; and Barbara Greenman of the University of Colorado libraries for her help with the ever-changing world of electronic information. I am also indebted to my family, for their patience, and especially my son, James Mack, who helped with many technical computer-associated aspects of this manuscript.

Patty L. Feist

Table of Contents

Introduction to the Organic Laboratory

1 Safety in the Laboratory

Organic chemistry is an experimental science. Our understanding of organic chemistry is mainly the result of laboratory observation and testing. For this reason, the laboratory is an important part of a student's education in organic chemistry.

Because of the nature of organic compounds, the organic chemistry laboratory is generally more dangerous than the inorganic chemistry laboratory. Many organic compounds are volatile and flammable. Some can cause chemical burns; many are toxic. Some can cause lung damage, some can lead to cirrhosis of the liver, and some are *carcinogenic* (cancer causing). Yet organic chemists generally live as long as the rest of the population because they have learned to be careful. When working in an organic chemistry laboratory, you must always think in terms of safety.

Summary of Safety Rules

It may happen that you are confronted with a laboratory accident and cannot remember exactly what to do. In such a situation, just remember the following:

- In the case of a spill: WASH!
- In the case of a fire: GET OUT!

In either case, your instructor or someone in a calmer frame of mind can then decide how to handle the situation.

Laboratory supervisors must ensure that students know and follow established safety rules, have and know how to use emergency equipment, and have information on special hazards in the laboratory. Of course, this is simple common sense, but it is also mandated by a Federal Laboratory Safety Standard: OSHA 29 CFR 1910.1450. This standard states that any laboratory in which hazardous chemicals are handled should have a written safety plan, known as the Chemical Hygiene Plan, or CHP. A CHP addresses specific hazards in the laboratory and procedures for managing them. The student, however, must accept responsibility for carrying out his or her own work in accordance with good safety practices and should be prepared in advance for possible accidents by knowing *what* emergency aids are available and *how* to use them if necessary.

A. Personal Safety

(1) Using Common Sense

Most laboratory safety precautions are nothing more than common sense. The laboratory is not the place for horseplay. Do not work alone in the laboratory. Do not perform unauthorized experiments. Do not sniff, inhale, touch, or taste organic compounds, and do not pipet them by mouth. Wipe up all spilled chemicals immediately according to the instructions of your laboratory instructor. Do not place hazardous chemicals in the waste basket or down the drain; instead, dispose of them according to the waste chemical policies in your laboratory. Position and clamp all apparatus carefully to prevent it from falling over. Do not leave broken glass laying around where someone else can get cut by it, and do not put it in the waste basket for an unsuspecting custodian.

When working in the laboratory, wear suitable clothing. Long pants and a shirt with long sleeves, plus a rubber lab apron or cloth lab coat, are ideal. Do not wear your best clothing—laboratory attire usually acquires many small holes from acid splatters and may also develop a distinctive aroma. Loose sleeves should be rolled up, or they might sweep flasks from the laboratory bench; loose sleeves present the added hazard of easily catching on fire. Long hair should be tied back. Broken glass sometimes litters the floor of a laboratory; therefore, always wear closed-toe shoes. Sandals are inadequate because they do not protect the feet from spills. Because of the danger of chemical contamination, food and drink have no place in the laboratory. Wear thick rubber gloves when handling chemicals. Wash your hands frequently, and always wash them before leaving the laboratory, even to go to the restroom.

Because of the danger of fires, smoking is prohibited in laboratories. On the first day of class, familiarize yourself with the locations of the fire extinguishers, fire blanket, eyewash station, and safety shower. Know the safe exits in case of an emergency. If the fire alarm sounds: Get out!

Use electrical equipment properly to prevent electrical shock. If a cord or plug is obviously damaged, do not use it. Keep water away from all electrical equipment.

Some laboratory procedures are associated with special hazards, for instance, the use of compressed gasses, reduced pressures, or particularly dangerous chemicals. Always read the safety notes or precautions in the procedure section of your laboratory manual *before* you begin an experiment, and think about safety as you carry out each step of a reaction.

(2) Eye Protection

Chemical splashes and flying broken glassware can lead to blindness; therefore, it is imperative that you wear eye protection in the laboratory. Chemical splash goggles, or simply **safety goggles**, afford the best protection since they fit snugly around the eyes to prevent chemicals from running down the forehead. These goggles also protect against the impact of broken glass. When trying on a pair of safety goggles, find a pair that fits well, has ventilation to prevent fogging, and allows for good peripheral vision. **Safety glasses** may also be worn (according to the policy at your academic institution), but they do not provide as good a level of protection for your eyes as goggles. Either type of eye protection must meet the basic standards of the American National Standard Practice, or ANSI.* When you purchase a pair of goggles or glasses, look for "ANSI Z87" on the box or imprinted on the goggles.

Wear eye protection *at all times*, even if you are merely adding notes to your laboratory notebook or washing dishes. You could be an innocent victim of your lab partner's mistake, who might inadvertently splash a corrosive chemical in your direction. In the case of particularly hazardous manipulations, you should wear a **full-face shield** (similar to a welder's face shield). Your instructor will tell you when this is necessary.

Contact lenses should not be worn, even under safety goggles because contact lenses cannot always be removed quickly if a chemical gets into the eye. A person administering first aid by washing your eye might not even realize that you are wearing contact lenses. In addition, "soft" contact lenses can absorb harmful vapors. If contact lenses are absolutely necessary, properly fitted goggles must be worn. Also, inform your laboratory instructor and neighbors that you are wearing contact lenses.

(3) Other Personal Protection

Personal protection equipment (PPE) includes eye and face protection (see above), clothing, gloves, and respirators. The type of PPE recommended for handling a particular chemical is often specified on the manufacturer's label. In the student laboratory, lab coats or aprons are usually sufficient to protect your skin and clothing. In certain industrial situations, full suits or overalls or even boots might be required. Protection against chemical vapors is provided by handling chemicals in fume hoods. A respirator can be used if a hood is not available or when you handle very toxic chemicals. Respirators come in many different types and styles, and special training is required for them to be used properly.

Gloves are often advised in student laboratories, but be aware that they vary greatly in their ability to protect your hands from contamination. The amount of protection a particular glove affords depends on the material and method of manufacture of the glove, the thickness of the glove, and the chemical it contacts. Thick rubber, neoprene, or nitrile gloves provide good protection from most chemicals although they are bulky and may make handling glassware and equipment awkward. Thin disposable Latex gloves, or surgical gloves, do not interfere with your ability to handle glassware, but they are permeable to most organic solvents. If you spill a solvent on these gloves while you are wearing them, you must remove them immediately. Disposable Latex gloves thus have a limited use in the organic laboratory. Some types of thin nitrile gloves do afford reasonable protection from organic solvents while still allowing good flexibility. Check with your laboratory instructor if you are unsure of the properties of your lab gloves.

---

* ANSI Z87.1, as mandated in OSHA 29 CFR 1910.133

B. Laboratory Accidents

(1) Chemicals in the Eyes

If a chemical does get into your eye, flush it with gently flowing water for 15 minutes. Do not try to neutralize an acid or base in the eye. Because of the natural tendency for the eyelids to shut when something is in the eye, *they must be held open during the washing*. If there is no eyewash fountain in the laboratory, a piece of rubber tubing attached to a faucet is a good substitute. Do not take time to put together a fountain if you have something in your eye, however! Either splash your eye (held open) with water from the faucet immediately or lie down on the floor and have someone else pour a gentle stream of water into your eye. *Time* is important. The sooner you can wash a chemical out of your eye, the less the damage will be.

After the eye has been flushed, medical treatment is strongly advised. For any corrosive chemical, such as sodium hydroxide, prompt medical attention is imperative.

(2) Chemical Burns

Any chemical (whether water-soluble or not) spilled onto the skin should be washed off immediately with soap and water. The detergent action of the soap and the mechanical action of washing remove most substances, even insoluble ones. If the chemical is a strong acid or base, rinse the splashed area of the skin with *lots and lots of cool water*. Strong acids on the skin usually cause a painful stinging. Strong bases usually do not cause pain, but they are extremely harmful to tissue. Always wash carefully after using a strong base.

If chemicals are spilled on a large area of the body, wash them off in the safety shower. If the chemicals are corrosive or can be absorbed through the skin, remove contaminated clothing so that the skin can be flushed thoroughly. If chemical burns result, the victim should seek medical attention.

(3) Heat Burns

Minor burns from hot flasks, glass tubing, and the like are not uncommon occurrences in the laboratory. The only treatment needed for a very minor burn is holding it under cold water for 5–10 minutes. To prevent minor burns, keep a pair of inexpensive, loose-fitting cotton gloves in your laboratory locker to use when you must handle hot beakers, tubing, or flasks.

A person with a serious burn, as from burned clothing, is likely to go into shock. Some symptoms of shock are clammy, pale skin; weakness; rapid but weak pulse; rapid, irregular breathing; restlessness; and unusual thirst. In the case of shock, get the person to lie down on the floor and keep him or her warm with a fire blanket or coat. Summon an ambulance immediately. Except to extinguish flames or to remove harmful chemicals, do not wash a serious burn and do not apply any ointments. However, cold compresses on a burned area will help dissipate heat.

(4) Cuts

Minor cuts from broken glassware are another common occurrence in the laboratory. These cuts should be flushed thoroughly with cold water to remove any chemicals or slivers of glass. A pressure bandage can be used to stop any bleeding.

Major cuts and heavy bleeding are a more serious matter. The injured person should lie down and be kept warm in case of shock. A pressure bandage (such as a folded, clean dish towel) should be applied over the wound and the injured area elevated slightly, if possible. An ambulance should be called immediately.

The use of a tourniquet is no longer advised. Experience has shown that cutting off all circulation to a limb may result in gangrene.

(5) Inhalation of Toxic Substances

A person who has inhaled vapors of an irritating or toxic substance should be moved immediately to fresh air. If breathing stops, administer artificial respiration and call an emergency medical vehicle.

C. Laboratory Fires

Your laboratory should have an Emergency Evacuation Plan (EEP) posted in each lab room. It is your responsibility to study this plan and know the quickest and safest way to exit the building in the event of a fire or other emergency. The best protection from fires, however, is preventing them in the first place.

(1) Avoiding Fires

Most fires in the laboratory can be prevented by the use of common sense. Before lighting a match or burner, check the area for flammable solvents. Solvent fumes are heavier than air and can travel along a benchtop or a drainage trough in the bench. These heavy, flammable fumes can remain in sinks or waste baskets for days. While it is indeed true that a flammable solvent should not have been discarded in the sink or waste basket, it is always possible that some inconsiderate fool has done so. Therefore, do not discard hot matches, even if extinguished, or any other hot substances in sinks or waste baskets.

Whenever you use a flammable solvent, extinguish all flames in the vicinity beforehand. Always cap solvent bottles when not actually in use. Do not boil away flammable solvents from a mixture except in the fume hood. Place solvent-soaked filter paper in the fume hood to dry before discarding it in a waste container. If you spill solvent on the benchtop or floor, clean it up immediately with paper towels (or a spill pillow) and then place the towels in the fume hood to dry.

Avoid spilling a flammable solvent on a hot surface, such as a hot plate or heating mantle, because it might catch fire.

(2) Extinguishing Fires

In case of even a small fire, tell your neighbors to leave the area and notify the instructor. A fire confined to a flask or beaker can be smothered with a watch glass or a large beaker placed over the flaming vessel. (Try not to drop a flaming flask—this will splatter burning liquid and glass over the area.) All burners in the vicinity of a fire should be extinguished, and all containers of flammable materials should be removed to a safe place in case the fire spreads.

For all but the smallest fire, the laboratory should be cleared of people. It is better to say loudly, "Clear the room!" than to scream "Fire!" in a panicky voice. If you *hear* such a shout, do not stand around to see what is happening, but stop whatever you are doing and walk immediately and purposefully toward the nearest clear exit, as directed in the Emergency Evacuation Plan.

If you need to extinguish a laboratory fire, use a *carbon dioxide* or a *dry powder fire extinguisher.* Water should never be used on a chemical fire. Many organic solvents float on water; therefore, water may serve only to spread a chemical fire. Some substances, like sodium metal, explode on contact with water.

If a fire extinguisher is needed, it is best to clear the laboratory and allow the instructor to handle the extinguisher. Even so, you should acquaint yourself with the location, type, and operation of the fire extinguishers on the first day of class. Inspect the fire extinguishers and note which letter, A-D, is on the unit. These letters refer to the class of fire for which the extinguisher is intended (see Table 1). Find the sealing wire (indicating that the extinguisher is fully charged) and the pin that is used to break this sealing wire when the extinguisher is needed.

Table 1 Classifications of fires.

Class of fire	*Substance that is burning*
A	ordinary combustibles (paper, wood, textiles, rubbish)
B	flammable liquids
C	electrical equipment
D	burning metals

Fire extinguishers usually spray their contents with great force. To avoid blowing flaming liquid and broken glass around the room, aim toward the base and to the side of any burning equipment, not directly toward the fire. Once a fire extinguisher has been used, it will need recharging before it is again operable. Therefore, any use of a fire extinguisher must be reported to the instructor.

(3) Extinguishing Burning Clothing

If your clothing catches fire, walk to the shower if it is close by. Never run, since this only fans the flames. If the shower is not near, lie down and roll to extinguish the flames. Rolling a person in a fire blanket can be used to smother the flames quickly. The rolling motion is important because a fire can still burn under the blanket. Wet towels can also be used to extinguish burning clothing. Whichever method is used, call for help immediately.

A burned person should be treated for shock (kept quiet, laying down with feet elevated if feasible, and warm). Medical attention should be sought.

D. Handling Chemicals

The chemicals that you use in student teaching laboratories are chosen by the laboratory supervisors as the safest chemicals that can be used in each situation. The experiments themselves are also chosen for safety. Still, organic chemistry laboratory techniques cannot be learned without the use of potentially hazardous chemicals.

Advance planning is one of the best ways to minimize accidental exposure to chemicals. Before you come to lab, look up the hazards of each chemical that you will be using in the experiment. Think about what would happen if you spilled each chemical on the benchtop, a hot surface, or yourself. Know the potential hazards so that you can take appropriate safety precautions.

The use of chemicals in academic chemistry laboratories falls under the guidelines of the Occupational Safety and Health Administration (OSHA). Briefly, students have the right to know what chemicals are stored in the laboratories and the hazards of those chemicals. A Material Safety Data Sheet, or **MSDS**, must be kept in the laboratory for each chemical used in the laboratory course. "Right to Know" laws and MSDS's are discussed in Appendix III (see p. 214).

(1) Acids and Bases

If you splash an acid or strong base on your skin, wash with copious amounts of water, as described in the section on chemical burns. To prevent acid splatters, always add concentrated sulfuric acid (H_2SO_4) to water (never add water to concentrated sulfuric acid). Concentrated sulfuric acid should be added to ice water or crushed ice because of the heat generated by the mixing. Concentrated hydrochloric acid (HCl) and glacial acetic acid (CH_3CO_2H) present the added hazard of extremely irritating vapors. These two acids should be used only in the fume hood. Do not pour acids or bases down the drain; instead, dispose of them as your laboratory instructor directs.

Sodium hydroxide (NaOH) is caustic and can eat away tissue. As a solid (usually pellets), it is deliquescent; a pellet that is dropped and ignored will form a dangerous pool of concentrated NaOH. Pick up spilled pellets while wearing gloves or by using a piece of paper, neutralize them, and place them in a disposal container as directed by your laboratory instructor.

Aqueous ammonia ("ammonium hydroxide") emits ammonia (NH_3) vapors and thus should be used only in the fume hood.

(2) Solvents

Organic solvents present the double hazard of flammability and toxicity (both short-term and cumulative). *Diethyl ether* ($CH_3CH_2OCH_2CH_3$) and *petroleum ether* (a mixture of alkanes) are both very volatile (have low boiling points) and extremely flammable. These two solvents should never be used in the vicinity of a flame, and they should be boiled only in the hood. *Benzene* (C_6H_6) is flammable and also toxic. It can be absorbed through the skin, and long-term exposure is thought to cause cancer. Benzene should be used as a solvent only when absolutely necessary (and then handled with great care to avoid inhalation, splashes on the skin, or fire). In most cases *toluene* can be substituted for benzene. Although toluene is flammable, it is less toxic than benzene.

Most halogenated hydrocarbons, such as *carbon tetrachloride* (CCl_4), *chloroform* ($CHCl_3$), and *methylene chloride* (CH_2Cl_2), are toxic, and some are carcinogenic. Halogenated hydrocarbons tend to accumulate in the fatty tissues in living systems instead of being detoxified and excreted, as most poisons are. In repeated small doses, they are associated with chronic poisoning and damage to the liver and kidneys. If either carbon tetrachloride or chloroform must be used, it should be handled in the fume hood.

Because of the dangers inherent with all organic solvents, they should always be handled with respect. Solvent vapors should not be inhaled, and solvents should never be tasted or poured on the skin. Wash any splashes on your skin immediately with soap and water. Keep solvent bottles tightly capped. Always heed precautions to avoid fires. Never put solvents down the drain.

(3) Disposal of Chemicals

The protocols for disposing of chemical wastes from classroom laboratories vary from institution to institution. The best rule of thumb is: "Nothing down the drain, nothing in the waste basket." Even solvent-soaked filter papers must be dried in the fume hood before being placed in the waste basket.

The laboratory personnel will provide containers for the wastes. These wastes will be segregated according to reactivity and compatibility. Some chemical wastes are harmless to the environment; however, it is up to the trained laboratory personnel to decide which chemicals can be disposed of in local waste treatment facilities. According to current government regulations, the wastes that are hazardous are packaged and sent to treatment facilities, where they are usually incinerated. The cost of packaging, transporting, and treating hazardous waste is often more than the cost of purchasing the chemical in the first place.

Before you start an experiment, be aware of the disposal procedures that apply to the wastes that will be generated by the experiment. This information should be part of the prelaboratory write-up of the experiment.

It is the responsibility of the chemist to be aware of and to obey all applicable disposal rules and regulations. If you don't know, ask. Ignorance is not an excuse.

2 The Laboratory Notebook

A laboratory notebook serves several purposes. The first is for your own reference. You may think that you will remember everything you do and see in the laboratory; however, many details, such as melting points, boiling points, and weights, are easily forgotten. It is far better to record these details in a well-organized notebook than to try to memorize them. An approved notebook should be used, not little slips of paper, which seem to flutter away when your back is turned.

Another purpose of a laboratory notebook is to allow someone else to review your work or repeat it *exactly*. This facet of experimental work is necessary in research, and it is also necessary in a student laboratory. If a particular experiment does not work well for you, the instructor would want to know why. Your detailed written procedure and observations can give clues to experimental failures.

A research laboratory notebook is also important to help establish the validity of patent claims. Each page or experiment in the notebook must be numbered and dated. If the project is especially promising, the researcher will have his or her notebook pages signed by witnesses. This procedure is usually unnecessary in a student laboratory.

A. The Correct Notebook

The correct notebook for the laboratory is a hardcover, bound book containing lined pages. These are available at bookstores and stationery stores. A loose-leaf or spiral notebook is not satisfactory because pages are easily removed and lost. A separate notebook should be used for each laboratory course.

B. Keeping the Notebook

If the pages in the notebook are not numbered, number them before using the book. Write your name and laboratory section number on the cover of your notebook. It is also advisable to put your address or telephone number on the book in case it is lost.

Leave two blank pages at the front of the book for a table of contents. Then, enter experiments consecutively in ink; use permanent ink, because your book will become splashed and stained with use.

Use only right-hand pages for writing up experiments. At the top of the page, write the date on which the experiment is performed. As you go along, leave plenty of space for notes that you might want to insert later. The empty left-hand pages may be used for calculations and jottings. If you will be running a distillation or determining more data for a particular experiment, be sure to leave blank pages as necessary before writing the procedure for the next experiment. If you make errors, do not rip out the page. Instead, line out errors (or draw an "X" over the entire page) and go on.

C. Entering Experiments

Each experiment in the notebook should contain the following information, along with any additional material required by your instructor.

(1) Title
(2) Balanced chemical equation
(3) Physical data
(4) Safety and disposal information
(5) Procedure outline
(6) Observations
(7) Conclusions

These items are described in detail next. Figure 1 shows a typical notebook page for the start of an experiment and also demonstrates the calculations you may have to perform.

(1) The **title** of the experiment should briefly describe the experiment and should also contain the experiment number (and other reference, where appropriate). If the experimental objective is not clear from the title, a concise statement describing the experiment should be included.

Experiment 10.1

Synthesis of 1-Bromobutane from 1-Butanol

$$CH_3CH_2CH_2CH_2OH + NaBr + H_2SO_4 \longrightarrow CH_3CH_2CH_2CH_2Br + NaHSO_4 + H_2O$$

	1-butanol	sodium bromide	sulfuric acid	1-bromobutane
MW:	74.12	102.90	98.08	137.03
weight:	18.5 g	30.0 g	25.0 mL (46.0 g)	34.2 g (theory)
moles:	0.250	0.292	0.469	0.250 (theory)

calculation of numbers of moles of reactants:

for 1-butanol: $\dfrac{18.5\ g}{74.12\ g/mol} = 0.250\ mol$

for NaBr: $\dfrac{30.0\ g}{102.90\ g/mol} = 0.292\ mol$

for H_2SO_4: $\dfrac{46\ g}{98.08\ g/mol} = 0.469\ mol$

calculation of theoretical yield of product:

for 1-bromobutane: $0.250\ mol \times 137.03\ g/mol = 34.2\ g$

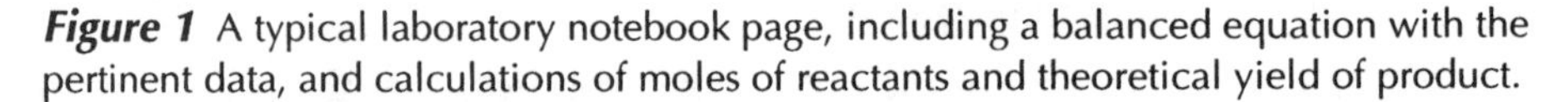

Figure 1 A typical laboratory notebook page, including a balanced equation with the pertinent data, and calculations of moles of reactants and theoretical yield of product.

(2) The **balanced chemical equation** should show the formulas and names of the reactants and products. The molecular weight, the actual weight used, and the number of moles should be written under the name of each reactant.

The molecular weight and the theoretical yield should be placed under the name of each organic product. The **theoretical yield** of a product may be calculated from its molecular weight and the number of moles of the **limiting reagent** (the reactant present in shortest supply). In the example in Figure 1, NaBr and H_2SO_4 are present in excess. Therefore, 1-butanol is the limiting reagent. In the example, the maximum number of moles of product that could be obtained from 0.250 mol of 1-butanol is 0.250 mol, or 34.2 g, as shown in the last calculation in Figure 1. Appendix I describes yield calculations in more detail.

If the experiment is not a reaction but is an isolation or purification experiment, then the formulas, names, and so on, of the compounds in question should be written out.

(3) The **physical data** for the reactants and products must be included. Pertinent physical properties such as boiling and melting points can be listed under the equation (under the molecular weight), or you can include this data in tabular form. The physical properties of organic compounds can be found in the MSDS, in printed reference texts, and on several Web sites (see Technique 17, p. 198).

Include the hazard information for each chemical used in the experiment. This information can be found either in the same reference handbook or resource as the physical data or in references that specialize in hazard information (see Appendix III, p. 213).

You should include the source of your physical and hazard data in your notebook.

(4) **Safety and disposal information** will most likely be provided in your experiment manual or by the laboratory instructor. If it is not, find the safety and disposal information on the MSDS. You should verify that your disposal procedures are in accordance with the rules established for your laboratory.

(5) The **actual procedure** that will be used should be outlined *in detail.* The purpose of the outline is to provide an overview of what you will be doing in the laboratory, thus allowing you to organize your time efficiently. (Do not waste your time by copying the procedure word for word. You can always refer back to the original procedure if necessary.)

A **flow diagram**, showing how the product will be isolated from by-products or unreacted starting material, should be included. Figure 2 shows a portion of such a diagram.

(6) **Observations** should be recorded in your notebook *as you do the experiment*. Examples of observations might be, "The ethanol solution was yellow," or "Approx. 5 mL of the reaction mixture was spilled and lost while being transferred to a separatory funnel." It is better to include too many observations than too few; however, use common sense. Entering the fact that you went to the storeroom to get a beaker wastes both your time and the time of anyone who reads your notebook. On the other hand, failure to record a weight or a physical constant may also waste your time—you may have to repeat the experiment.

(7) **Conclusions** of your experiment will usually include the weight and physical properties (such as melting point or boiling point) of the isolated and purified product plus the **percent yield**. This percent yield is the actual percent of the theoretical yield that you obtained.

$$\text{percent yield} = \frac{\text{actual yield in g}}{\text{theoretical yield in g}}$$

Thus, you might write in your notebook:

Yield: 5.3 g (61%) of benzoic acid, mp 117–119°
mp of an authentic sample of benzoic acid, 119.5–120°
mixed mp of product with authentic sample, 118.5–119°

In some experiments, you may have to prove the structure of a compound. In these cases, your conclusion should include all the supporting data used in the structure proof. These data may consist of physical constants, spectroscopic information, and chemical reactivity.

You may be required to turn in reaction products to your instructor. Your instructor will provide sample bottles or will suggest other containers. Any sample you turn in should be labeled clearly with the compound's name, weight, and pertinent physical constants. Your name and laboratory section number must also be included. Use waterproof ink for your label. Your instructor may also want you to cover the label with transparent tape.

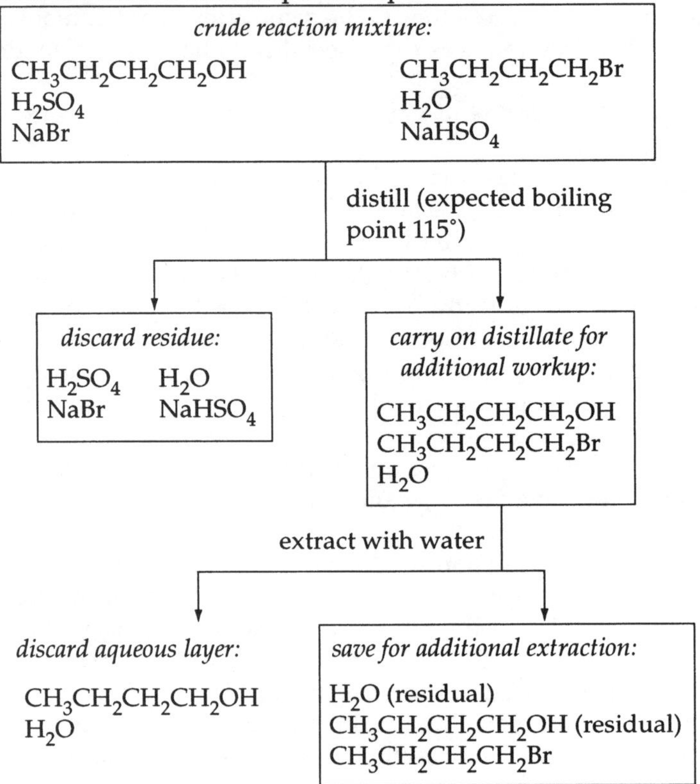

Figure 2 A partial flow diagram for the isolation of 1-bromobutane.

3 Laboratory Equipment

A. Glassware

Figure 3 shows the glassware found in a typical student locker. Your own laboratory may not supply every item shown or it might contain items not shown.

Note the ground-glass joints on the round-bottom flask, condenser, and so on. These joints are ground to a *standard-taper.* The symbol for standard-taper glassware is shown below:

$ standard-taper symbol*

The size of a standard-taper joint is identified by a pair of numbers, such as 19/22, 14/20, or 24/40. In each pair, the first number refers to the outside diameter of the inner joint at its widest part, and the second number refers to the length of the joint. Any 19/22 inner joint will fit any 19/22 outer joint; therefore, glassware of the same standard taper is interchangeable. The advantages of ground-glass joints are that they provide a good seal between two pieces of equipment and that a laboratory setup can be assembled quickly.

One disadvantage of ground-glass joints is their tendency to stick, or freeze. A film of hydrocarbon or silicone grease on these joints is used to prevent sticking. When you assemble ground-glass equipment, grease each inner joint *lightly*, insert it into the outer joint, and rotate the joints to distribute the grease evenly. The glass stopcocks of separatory funnels or dropping funnels also require grease, but Teflon® stopcocks do not. The cleaning and storing of ground-glass equipment is discussed in Section 4, Cleaning Glassware (p. 18).

Equipment with ground-glass joints is expensive. If you break a piece of glassware containing a ground-glass joint, do not discard the joint unless your instructor tells you to do so. A competent glassblower can recycle the joints by sealing them on ordinary, less expensive flasks and condensers.

When you check the equipment in your locker, check each piece of glassware carefully for cracks and "stars" (small star-shaped cracks). Each time you use a piece of glassware, recheck it. Any cracked or starred glassware should be returned to the storeroom and replaced by undamaged ware. In some cases, a glassblower will be able to repair the cracks.

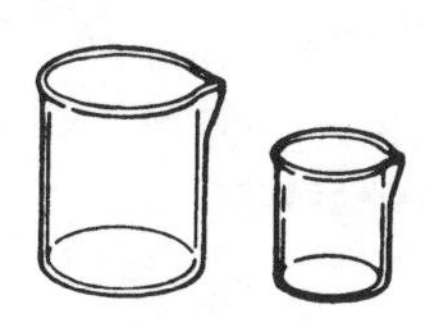

beaker, assorted sizes

funnel (conical)

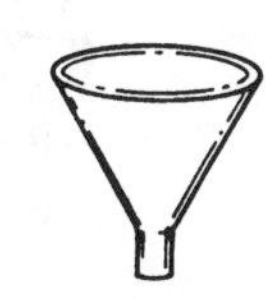

powder funnel
(with thick stem)

stemless funnel

Büchner funnel
(for vacuum filtration)

Hirsch funnel
(for vacuum filtration)

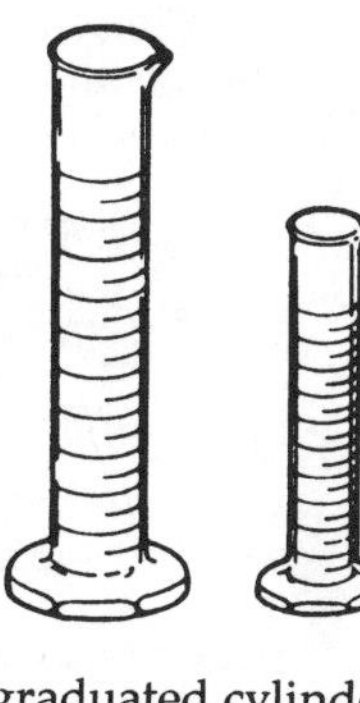

graduated cylinder,
assorted sizes

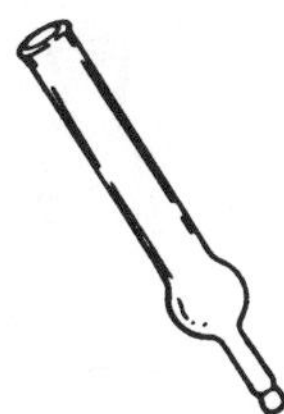

drying tube

watch glass

Figure 3 Typical glassware stored in a student locker.

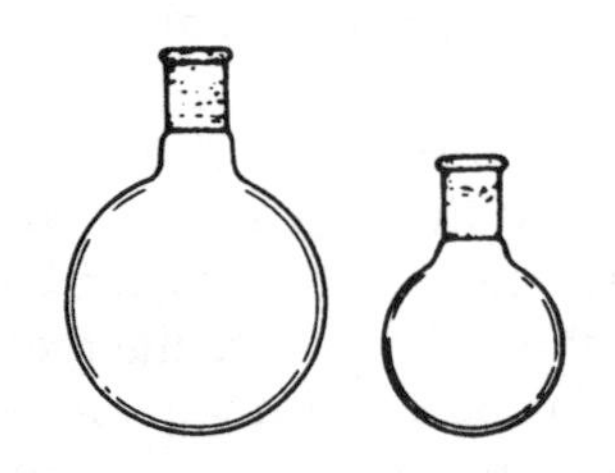

round-bottom flask, assorted sizes
(for reactions and distillations)

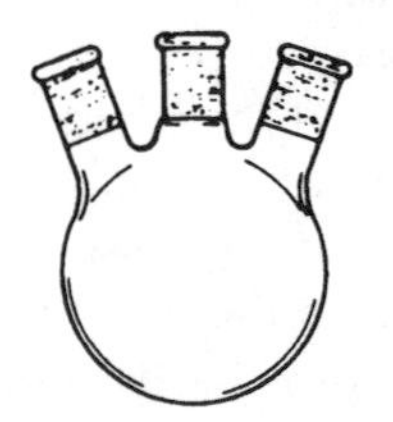

three-neck round-bottom flask
(used with a reflux condenser, stirrer, and dropping funnel)

separatory funnel
(for extractions)

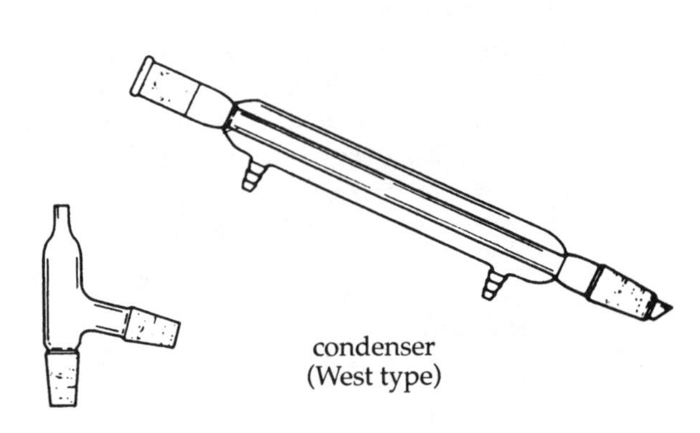

distillation head

condenser
(West type)

distillation adapter

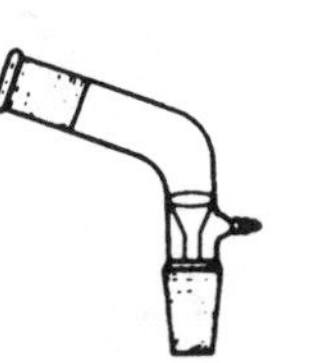

vacuum distillation adapter

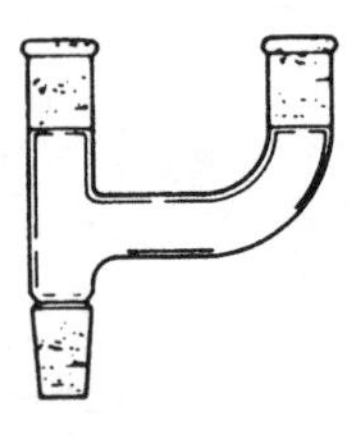

Claisen head
(for distillations and reaction assemblies)

dropper
(disposable pipet)

Erlenmeyer flask, assorted sizes

heavy-walled filter flask
(for vacuum filtration)

Figure 3 (continued)

B. Nonglass Equipment

Other useful items often found in a student locker are pictured in Figure 4. Spatulas are used for transferring solids. Clamps and metal rings support glassware. A Filtervac or neoprene adapter is used to attach a Büchner or Hirsch funnel to a filter flask. However, a one-holed rubber stopper that fits both funnel and flask gives a better seal and is easier to use.

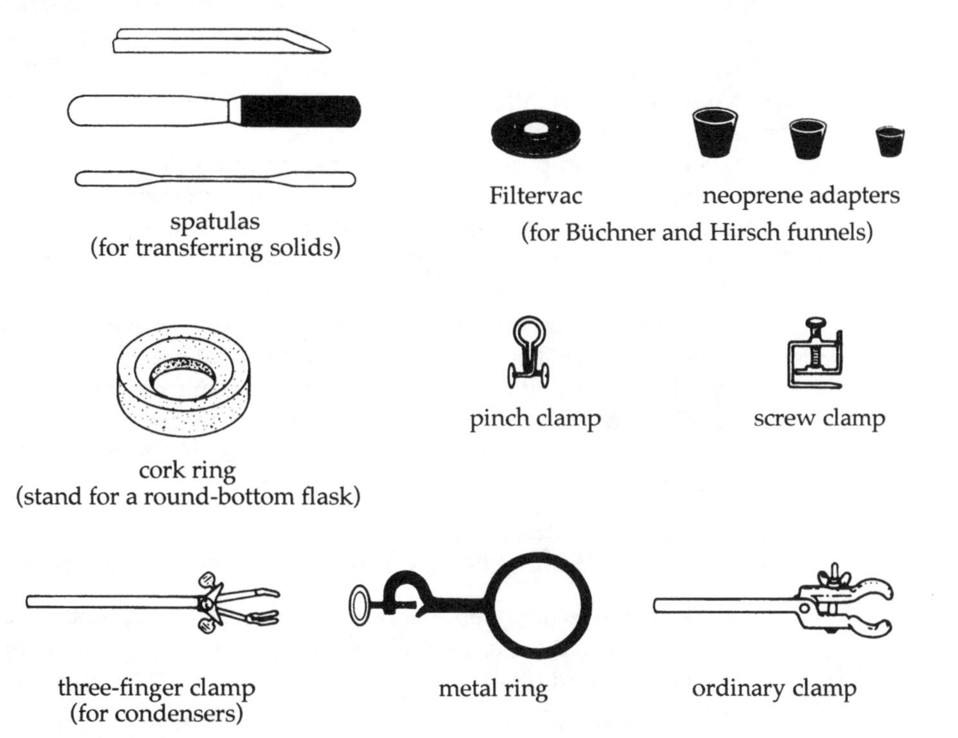

Figure 4 Hardware and other nonglass items typically found in a student locker.

C. Heating Equipment

Several devices for heating liquids in flasks are shown in Figure 5. The **steam bath** heats liquids to a maximum of about 90°C and is useful for heating low-boiling solvents, especially flammable ones. To clear water from the steam line, turn the steam on forcefully. Then turn the steam to low before placing a flask on it. (Too much steam will allow water to contaminate the contents of the flask, and its noise is irritating to others in the laboratory.) A round-bottom flask can be set into the appropriately sized ring of the steam bath, while an Erlenmeyer flask can be set on top of a ring smaller than itself. If all the rings are left on top of the steam bath, two or three small Erlenmeyer flasks can be warmed simultaneously.

Modern **hot plates** are safe for heating flammable solvents, as long as they do not develop very hot surfaces, have exposed electrical coils, or spark when the thermostat clicks on and off. Check with your laboratory instructor if you have any questions about the safety of the hot plates in the laboratory. Hot plates can be used in conjunction with a **sand bath** (simply a container of sand placed on top of the hot plate) to heat glass reaction containers that do not have a flat bottom. Sand baths are common in laboratories using microscale glassware.

A **heating mantle** is used for heating the contents of a round-bottom flask. With a heating mantle, the flask becomes hotter than its contents. To avoid decomposition of material splashed on the hot glass, never use a heating mantle with a nearly empty flask. Heating mantles are available in assorted sizes. A soft, glass coil mantle should fit a flask snugly. A hard, ceramic coil mantle need not fit a flask snugly and can thus accommodate more than one flask size. Before use, make sure that your heating mantle does not appear worn and that the heating element is not exposed. A worn heating mantle can cause a fire!

The electrical plug of a typical heating mantle will not fit into the standard wall socket, because heating mantles are not designed to operate at 110 volts. Instead, the mantle is plugged into a **variable transformer**, or **rheostat** (Variac®, Powerstat®, Powermite®), which is used to adjust the voltage and thus the temperature. Figure 5 shows a transformer of this type.

Bunsen burners and microburners are used for bending glass tubing and for other glassworking. They may also be used to heat aqueous solutions and high-boiling liquids. The use of a wire gauze with a ceramic center is recommended to distribute the heat from a burner to a flask and to prevent localized overheating. Burners should be used with the utmost caution in the organic laboratory because of the ever-present vapors of organic solvents.

Safety Notes Summary: Heating Devices

- Do not use flammable solvents with hot plates that develop very hot surfaces, have exposed electrical coils, or spark when the thermostat clicks on and off.
- Do not use a heating mantle that appears worn or has the heating element exposed.
- Never light a burner (or even a match) if someone is using a flammable solvent nearby. (Check with your instructor if you are unsure whether it is safe to light a match.)
- Never use a burner to heat a flammable solvent.
- Never open a solvent bottle without first checking the vicinity for flames.

D. Community Equipment

You will share a **balance** with your classmates. Your instructor will demonstrate the operation of the balance(s) in your laboratory. A balance is a delicate instrument: Treat it with care, and always wipe up spills (liquid or solid) on or near the balance. Do *not* blow spilled powders off the balance pan—many chemicals that are ordinarily safe to handle are dangerous when inhaled.

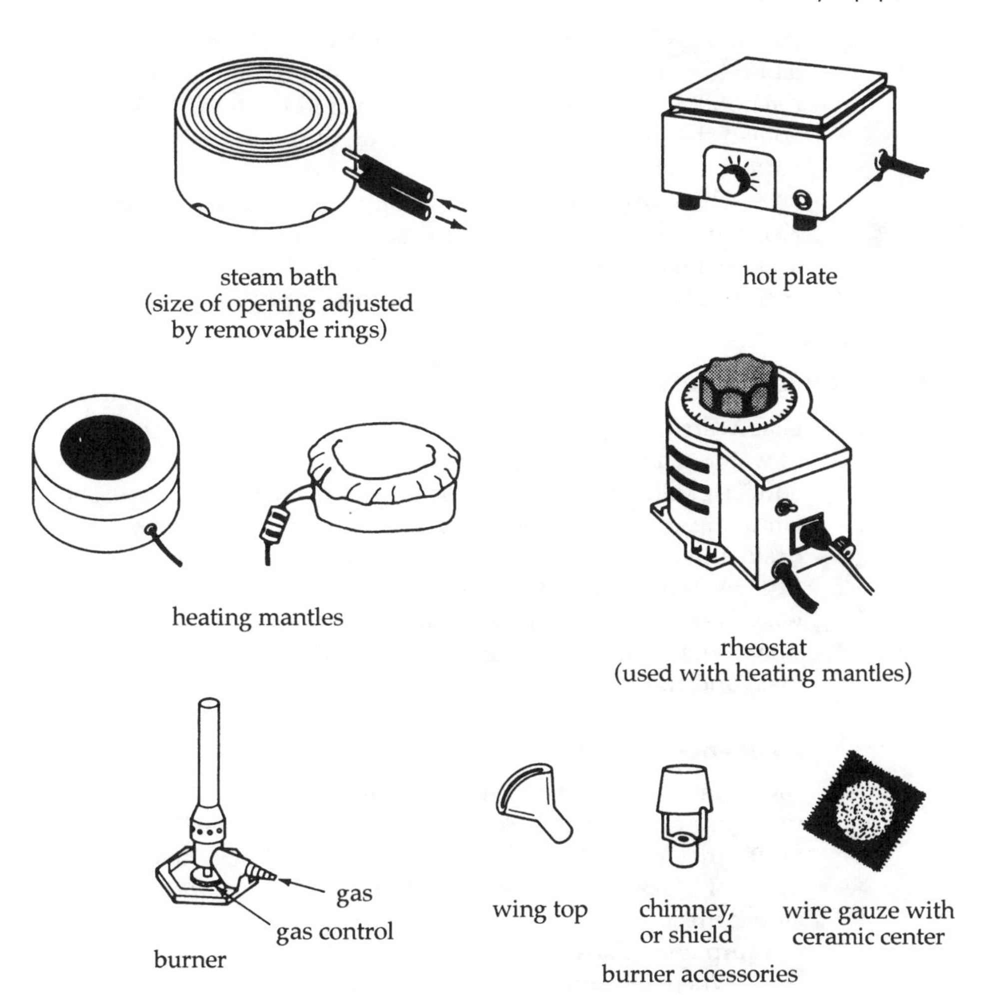

Figure 5 Some typical laboratory heating devices.

You may be instructed to use a **stir motor** when carrying out reactions (see Technique 14, p. 153). These devices have an enclosed, spinning magnet; if a Teflon® covered magnetic **spin bar** is placed in a glass flask above the unit, the bar spins, and any liquid in the flask is stirred. Another type of mechanical stirring device is an overhead stirrer, in which a paddle on a rod is lowered into a flask of liquid and the motor in the stirrer causes the paddle to spin.

You will probably be using analytical electronic equipment such as gas chromatographs and spectrometers in your laboratory course. These are also delicate instruments. Do not be a knob twister. Understand and follow your instructor's directions.

If you encounter problems with any of the laboratory's electrical or electronic equipment, notify your instructor.

The **fume hood** should be operating whenever someone is working in the laboratory. Make sure the glass door of the fume hood is kept at or below the arrow marked for proper air flow. It is recommended that highly flammable or toxic materials *always* be handled in the hood. Unfortunately, space limitations in a student laboratory do not always permit this. Because of the nature of the chemicals used in the hood, it is imperative that the bench in the hood be kept clean. If you find some spilled chemical of unknown origin on the hood bench, clean it up as if it were toxic, corrosive, and flammable.

The sharing of other community equipment is a matter of safety and courtesy. Keep reagent bottles tightly capped. Return them to their proper spots promptly and do not take them to your workstation. Do not contaminate chemicals in reagent bottles. If you see that a bottle is almost empty, report the fact to your instructor.

4 Cleaning Glassware

Glassware for organic chemical reactions should be both *clean* and *dry.* The presence of water in a flask can ruin many experiments (except, of course, those performed in aqueous solution). For this reason, glassware should be washed as soon as you finish using it (certainly by the end of each laboratory period) and then allowed to drain and dry in your locker until the next laboratory period.

Another reason to wash your glassware promptly is that freshly dirtied glassware is easier to clean than glassware containing dried-out tars and gums. Furthermore, some compounds, like sodium hydroxide and potassium hydroxide, can etch glass and ruin ground-glass joints if left standing in flasks.

A. General Cleaning

Most glassware can be cleaned readily with strong industrial detergent or scouring powder and a bottle brush. The brush should be bent, if necessary, to reach the entire inside of a flask. The detergent should be rinsed out thoroughly, then the piece of glassware rinsed with a small amount of distilled water. A flask can be stored upside down on a crumpled towel in a beaker to drain and dry. Other pieces of equipment, like condensers, can be laid on their sides in the locker.

When absolutely necessary, glassware can be dried quickly by rinsing it with a small amount of acetone. (Because of the expense, some laboratories do not allow acetone to be used for rinsing flasks.) Never use expensive reagent-grade acetone for rinsing flasks! Use "wash acetone" instead. If it is used to rinse only water from a flask, wash acetone may be reused several times before it is discarded in the waste acetone container. Acetone, like any solvent, should be handled with care. It should not be poured on the skin, and its vapors should not be inhaled.

A flask rinsed in acetone will dry fairly quickly in the air. The drying process can be speeded up by placing a clean glass dropper, connected to a vacuum line by heavy-walled rubber tubing, in the drained flask. The vacuum will suck fresh air through the flask, sweeping acetone and water vapors into the vacuum line. Using a stream of air to dry an acetone-rinsed flask is not recommended. Compressed air is likely to contain droplets of water and oil that will contaminate the flask. Also, a noisy blast of air shot into a flask can startle other people working in the laboratory.

Never use a flame to dry a flask. If the flask contains water droplets, the flask may become unevenly heated and crack. If the flask contains solvents, the vapors may catch fire. A drying oven may be used to dry water-rinsed glassware but not glassware rinsed with a solvent. Before placing glassware in the oven, separate any ground-glass joints and remove stopcocks. Do not dry Teflon® stopcocks in a drying oven.

Ground-glass equipment should be dismantled and cleaned before it is placed in your locker. The joints should be kept free of chemicals and grit; otherwise, they may become stuck together, or *frozen*. A frozen joint can sometimes be unfrozen by rinsing the outer portion in hot water to cause it to expand. A *gentle* rap on the tabletop might also loosen it. There are some more sophisticated techniques for unfreezing joints; your instructor can advise you.

Glass stopcocks in separatory funnels are treated as other ground-glass joints and should be cleaned of hydrocarbon greases with a tissue dampened with acetone and stored separately. (Silicone greases should be carefully removed with dichloromethane.) If you would rather keep your glass stopcock in the separatory funnel, regrease it before replacing it. Teflon® stopcocks, which do not need greasing, should be cleaned and then replaced loosely in the joint until they are used again.

B. Hard-to-Clean Flasks

Tars and gums, which are large organic molecules called **polymers**, are formed when a large number of organic molecules react to yield very long chains or three-dimensional molecular networks. These substances are not soluble in water. Therefore, they should be scraped out of a flask with a metal spatula and discarded in a waste crock, not in the sink. Acetone is a good solvent for most organic compounds and is often useful for dissolving remnants of tars and other organic residues from dirty flasks. In some cases, swirling a small amount of acetone in a flask will dissolve the organic residues. In other cases, several hours of soaking will be necessary.

If you cannot get a flask clean by the methods mentioned above, check with your laboratory instructor or the storeroom personnel. Stronger cleaning agents are available, but these agents are hazardous and must be used only under the supervision of trained personnel.

Additional Resources

A bibliography of printed safety resources is published on the Brooks/Cole Web site, as well as additional problems pertinent to safety, laboratory notebooks, and laboratory equipment. The Web site also maintains current links to Internet resources for OSHA standards, first aid and emergency procedures, disposal of chemicals, and other useful information. Please visit www.brookscole.com.

Problems

1 Give reasons for the following safety rules.
 (a) Contact lenses should not be worn in the laboratory.
 (b) A chemical spill on the skin should be washed off with water, not with solvent.
 (c) Solvents are not to be poured down the sink.
 (d) Water should not be used to extinguish laboratory fires.
 (e) To dilute concentrated sulfuric acid, we pour it onto ice instead of simply mixing it with water.
 (f) Broken glassware should be picked up immediately and put in the designated container.
 (g) Closed-toe shoes should be worn to laboratory.

2 What should you do in each of the following circumstances?
 (a) Your neighbor splashes a chemical into his or her eye.
 (b) A strong acid spills onto your hands.
 (c) You spill a large amount of diethyl ether on the bench top.
 (d) Your neighbor's clothing catches fire.
 (e) Your reaction flask catches fire.

3 Find the following items in your student laboratory:
 (a) Fire extinguishers.
 (b) Safety wash or eye wash.
 (c) EEP (Emergency Evacuation Plan).
 (d) MSDS's.
 (e) First aid kit.
 (f) Closest phone for use in emergency situations.

4 Why is it important to clean up any chemical that you spill in the laboratory? If you do not know the identity of the chemical, where should you put it?

5 Find the hazard information for each of the following compounds from three different sources, including one MSDS, one printed source, and one Web or online source (See Appendix III). Compare and contrast the relative dangers of working with each chemical.
 (a) Diethyl ether
 (b) Benzene
 (c) Methylene chloride
 (d) Ethanol
 (e) Benzophenone
 (f) Sodium chloride

6 Look up the physical properties of the following compounds:
(a) 1-Bromopentane
(b) Methanol
(c) Sodium bromide

7 Make the following conversions. (Note: you should have looked up the physical data for the compounds in a, b, and c for Problem 6.)
(a) 5.0 g $CH_3CH_2CH_2CH_2CH_2Br$ to moles
(b) 0.100 mol CH_3OH to grams
(c) 2.50 mol NaBr to grams
(d) 10.0 mL concd H_2SO_4 (physical properties inside front cover) to moles
(e) 0.30 mol H_2SO_4 to mL of 6N H_2SO_4

8 You need 45 mL of a 5% aqueous solution of $NaHCO_3$.
(a) What weight of $NaHCO_3$ is required?
(b) How much water will you add?

9 What volume of 50% NaOH is needed to prepare 25 mL of 2.0% NaOH?

10 Calculate the percent yield when:
(a) the theoretical yield of a product is 15.3 g and a student obtains 6.9 g
(b) the theoretical yield is 3.1 g and a student obtains 2.7 g

11 For each of the following reactions, (1) identify the limiting reagent, and (2) calculate the theoretical yield of the organic product. (Note: The equations as shown are not necessarily balanced.)

(a) $\underset{25.0\ g}{CH_3CO_2H} + \underset{10.0\ g}{NaOH} \longrightarrow CH_3CO_2Na + H_2O$

(b) $\underset{5.0\ g}{H_2NCH_2CH_2CH_2NH_2} + \underset{10.0\ mL}{HCl\ (12\ M)} \longrightarrow Cl^-H_3\overset{+}{N}CH_2CH_2CH_2\overset{+}{N}H_3Cl^-$

12 What would be the heat source (or heat sources) of choice for boiling each of the following solvents?
(a) Diethyl ether, $(CH_3CH_2)_2O$, bp 35°
(b) Water, bp 100°
(c) Ethanol, CH_3CH_2OH, bp 78°
(d) Acetone, $(CH_3)_2C{=}O$, bp 56°

13 Many student laboratories employ heating mantles for heating reaction mixtures. These devices should not be plugged directly into the electrical outlet. Why?

14 You open your laboratory drawer to prepare your glassware for an experiment. You find that your round-bottom flask is attached so firmly to your distillation adapter that you cannot take them apart.
(a) What should you have done when storing your glassware during the previous lab period?
(b) How can you get the two pieces of glassware apart?

15 At the end of a reaction, your glassware is covered with a tarry substance. If you use acetone to clean the tar from the glassware, can you place the glassware immediately in a drying oven? Why or why not?

16 You have a summer job as a laboratory technician at a local chemical manufacturing company. What information about safety plans and the chemicals you will be working with should your company or supervisor provide?

Crystallization

When a solid organic compound is prepared in the laboratory or isolated from a natural source, it is almost always impure. A simple technique for the purification of such a solid compound is **crystallization**. To carry out a crystallization, dissolve the compound in a minimum amount of hot solvent. If insoluble impurities are present, the hot solution is filtered. If the solution is contaminated with colored impurities, it may be treated with decolorizing charcoal and filtered. The hot, saturated solution is finally allowed to cool slowly so that the desired compound crystallizes at a moderate rate. When the crystals are fully formed, they are isolated from the **mother liquor** (the solution) by filtration.

If an extremely pure compound is desired, the filtered crystals may be crystallized again, often referred to as **recrystallization**. Of course, each crystallization results in some loss of the desired compound, which remains dissolved in the mother liquor along with the impurities.

Crystallization is the *slow* formation of a crystalline solid, as opposed to precipitation, which is the *rapid* formation of an amorphorous solid. If a hot, saturated solution is cooled too quickly, the compound may precipitate instead of crystallizing. A precipitated solute may contain many impurities trapped in the rapidly formed amorphous mass by entrainment. On the other hand, when a solution is allowed to crystallize slowly, impurities tend to be excluded from the growing crystal structure because the molecules in the crystal lattice are in equilibrium with the molecules in solution. Molecules unsuitable for the crystal lattice are likely to remain in the solution, and only the most suitable molecules are retained in the crystal structure. Because impurities are usually present in low concentration, they remain in solution even when the solution cools.

To understand why a slow and careful crystallization is preferable to a rapid precipitation, consider the mechanism of crystallization. Crystallization occurs in stages. As the hot, saturated solution cools, it becomes supersaturated; then crystal nuclei form. These nuclei often form on the walls of the container, at the liquid surface, or on a foreign body (such as a dust particle), because there is a greater probability of proper molecular association at these locations.

Once the crystal nuclei have been formed, additional molecules migrate to their surfaces by diffusion and join the crystal lattice. Because the molecules must migrate from the bulk of the solution to the growing crystal surface, the solution surrounding the crystal becomes less concentrated than the bulk of the solution. Also, crystal growth is usually exothermic. So the heat released from the growing crystal increases the solubility of the compound near the surface. For crystallization to continue, the concentration of solute at the crystal site must be increased and the heat must be dissipated. These processes occur by diffusion and take time. Premature chilling or agitation can increase the rate of crystal growth to the point at which a precipitate (an amorphous solid) forms. The purest crystals are obtained when crystallization occurs slowly from an undisturbed solution.

1.1 Solvents for Crystallization

The ideal solvent for the crystallization of a particular compound is one that

- does not react with the compound
- boils at a temperature below the compound's melting point
- dissolves a moderately large amount of the compound when hot
- dissolves only a small amount of the compound when cool
- is moderately volatile so that the final crystals can be dried readily
- is nontoxic, nonflammable, and inexpensive
- does not dissolve impurities when hot
- does dissolve impurities when cold

As you might guess, a solvent possessing *all* of these attributes does not exist. The primary consideration in choosing a solvent for crystallizing a compound is that the compound be moderately soluble in the hot solvent and less so in the cold solvent. Unfortunately, the solubility of a compound in a solvent cannot be predicted with accuracy. Most commonly, in the selection of a specific solvent for a specific compound, the solubility of the compound in various solvents is determined by trial and error. If the best solvent for crystallizing a compound is not known, small portions of the compound can be tested with a variety of likely solvents (see Section 1.3B, p. 35).

General guidelines for predicting solubilities based on the structures of organic compounds do exist. For example, an *alcohol*, a compound containing the hydroxyl (–OH) group as its functional group, may be soluble in water because it can form hydrogen bonds with water molecules. *Carboxylic acids* (compounds containing $-CO_2H$ groups) and *amines* (compounds containing $-NH_2$, –NHR, or $-NR_2$ groups) also can form hydrogen bonds and are also generally soluble in polar solvents such as water or alcohols.

As the amount of hydrocarbon in the compound increases, the compound's solubility in water will decrease, but it still may be soluble in an alcohol, such as ethanol. Compounds that are largely hydrocarbon in structure are not soluble in polar solvents because C–C and C–H bonds are not polar. For these compounds, we would choose a nonpolar solvent—for example, petroleum ether, which is a mixture of alkanes such as pentane, $CH_3(CH_2)_3CH_3$, and hexane, $CH_3(CH_2)_4CH_3$. Thus, in choosing crystallization solvents, chemists generally follow the rule of the thumb that **like dissolves like**. Table 1.1 lists some common crystallization solvents, arranged according to their polarities.

For well-known compounds, suitable crystallization solvents have already been determined. Procedures for experiments in laboratory textbooks and chemical journals usually designate the optimal crystallization solvent.

Ideally, a compound to be crystallized should be soluble in the hot solvent but insoluble in the cold solvent. When such a solvent cannot be found, a chemist may use a **solvent pair**. A solvent pair is simply two miscible liquids chosen so that one

Table 1.1 Some common crystallization solvents, listed in order of decreasing polarity.

Name	*Formula*	*Dielectric constant**	*bp (°C)*	*Comments*
water	H_2O	78.5	100	—
methanol	CH_3OH	32.6	65	flammable, toxic
ethanol (95%)	CH_3CH_2OH	24.3	78	flammable
acetone	$(CH_3)_2C{=}O$	20.7	56	flammable
methylene chloride†	CH_2Cl_2	9.1	40	toxic
ethyl acetate	$CH_3CO_2CH_2CH_3$	6.0	77	flammable
chloroform	$CHCl_3$	4.8	61	toxic
diethyl ether	$(CH_3CH_2)_2O$	4.3	35	highly flammable
toluene	$C_6H_5CH_3$	2.4	111	flammable
cyclohexane	C_6H_{12}	2.0	81	flammable
hexanes	C_6H_{14}	2.0	67–69	flammable
petroleum ether‡	C_nH_{2n+2}	~1.8	~30–60	flammable

* Dielectric constant (a measure of polarity) at about 20°–25°.
† Also known as dichloromethane.
‡ A mixture of alkanes boiling at various ranges as specified by the manufacturer. Ranges of 30°–60° are usually designated "low-boiling petroleum ether" and ranges of 60°–90° are generally designated "high-boiling petroleum ether." High-boiling petroleum ether is sometimes called "ligroin." Note that petroleum ether is a mixture of alkanes and not a true ether.

liquid dissolves the compound readily and the other does not. For example, many polar organic compounds are very soluble in ethanol but insoluble in water. To crystallize such a compound, dissolve it in a moderate amount of hot ethanol; then add water drop by drop until the solution becomes turbid (cloudy). Finally, add a few drops of ethanol to redissolve the precipitating compound. The resulting ethanol–water solution is a saturated solution and is allowed to cool slowly so that crystallization will occur. Table 1.2 lists some common solvent pairs.

Table 1.2 Some common solvent pairs for crystallization.

methanol–water	diethyl ether–methanol
ethanol–water	diethyl ether–acetone
acetone–water	diethyl ether–petroleum ether
benzene–hexanes	methanol–methylene chloride

1.2 Steps in Crystallization

Note: The steps discussed in the following paragraphs are summarized in chart form in Figure 1.3, p. 33.

(1) Dissolving the Compound

The first step in crystallization is dissolving the compound in a minimum amount of the appropriate hot solvent in an Erlenmeyer flask. Be sure to add boiling chips to the flask before you bring the solvent to a boil (see Technique 14, p. 155). An Erlenmeyer flask is used instead of a beaker or other container for several reasons. The solution is less likely to splash out and dust is less likely to get in. The sloping sides allow boiling solvent to condense and return to the solution and allow easy removal of crystals. Also, an Erlenmeyer flask can be corked and stored in your locker.

Pulverize a lumpy solid with a spatula or glass rod to dissolve it more rapidly. To ensure that a minimum amount of solvent is used, add the solvent a few milliliters at a time and heat the mixture with constant stirring or swirling. When almost all of the solid has dissolved, examine the solution and the bottom of the flask for insoluble impurities. If impurities are visible, do not add excess solvent in an attempt to dissolve them, but filter the hot solution (step 2, below). This hot filtration is not necessary and is, in fact, undesirable if the solution looks clear and clean. If the solution appears to be contaminated with colored impurities, decolorizing charcoal may be added at this time. The use of decolorizing charcoal is discussed in Section 1.3C, p. 36.

If impurities are not visible and if the solution is not contaminated with colored impurities, skip the next step and go directly to step 3, Crystallizing the Compound.

(2) Filtering Insoluble Impurities

Filtering a hot, saturated solution inevitably results in cooling and in evaporation of some of the solvent. Therefore, a premature crystallization of the compound on the filter and in the funnel may occur. A few precautions can minimize this premature crystallization.

To help prevent clogging in the funnel, choose a stemless funnel, a short-stemmed funnel, or a powder funnel. Preheat the funnel by placing boiling chips and a small amount of solvent in the receiving Erlenmeyer flask, resting the funnel on top, and heating. Alternatively, warm the funnel on the flask containing the hot solution to be filtered.

Before filtering, add a little extra solvent (about 5–10% of the total volume) to the solution, and keep the solution hot while preparing the filtration apparatus. Filter the hot solution through either filter paper or a plug of glass wool. If you are using filter paper, choose a porous paper. Filter paper is rated by its porosity, with higher numbers corresponding to lower porosity. For hot filtration, use Whatman's No. 1 or 2; do not use No. 5 or 6, which have slower filter speeds.

Fluted filter paper is preferred to folded filter paper, because the increased surface area of the fluted paper allows the filtration to proceed more rapidly. Figure 1.1 shows how to prepare a piece of fluted filter paper. Place the fluted filter paper in the warm funnel in the neck of the receiving flask. The funnel must be supported slightly away from the lip of the flask in order to prevent a liquid seal from blocking the flow of air and solvent fumes.

To pour the hot solution, wrap the hot flask in a towel or hold it in a clamp. Do not use a test tube clamp or tongs, because they do not have enough strength to hold the flask. Alternatively, use a pair of inexpensive cotton gloves.

During the filtration, keep both flasks hot on a steam bath or hot plate. To keep the solution hot, pour only small amounts into the filter paper (instead of filling the filter paper to the brim). If a flammable solvent and a hot plate are used, move the flasks away from the hot plate when pouring so that solvent vapors do not flow over the heating element.

If crystallization occurs in the funnel, you can often remove the crystals by heating the receiving flask to boiling with the funnel still on it. Solvent condensing in the funnel may dissolve the crystals and carry them back to the filtered solution. Alternatively, wash the solid into the flask with a little hot solvent.

After all of the hot solution has been filtered, wash the original flask with a small amount of hot solvent. Pour this solvent through the filter paper into the receiving flask to transfer the final traces of the desired compound. Two washings may be necessary; however, use a minimum amount of solvent.

The crystallization flask should contain a hot, clear, saturated solution of the compound. Boil away excess solvent at this time using boiling chips or a boiling stick to prevent bumping. (Remember to use the hood for a toxic or flammable solvent.) If the hot solution starts to crystallize, reheat it to dissolve the crystals. If so much solvent has evaporated that these crystals will not redissolve, add a small additional amount of solvent to the flask and bring the mixture to a boil.

(3) Crystallizing the Compound

Cover the flask containing the hot, saturated solution with a watch glass or inverted beaker to prevent solvent evaporation and dust contamination. Then set the flask aside where it can remain undisturbed (no jostling or bumping, which will induce precipitation rather than crystallization) for an hour or several hours. If the flask must sit for several days, allow it to cool to room temperature. Then stopper it with a cork (not a rubber stopper if an organic solvent was used) to prevent solvent evaporation.

Chilling the mixture in an ice-water bath after crystallization appears complete will increase the yield of crystals. Be sure to allow ample time for the final crystal growth to occur before chilling.

Sometimes a hot solution cools to room temperature with no crystallization occurring. In such a case, your first question should be, "Is the solution *supersaturated*?" Often, crystallization can be induced in a supersaturated solution by scratching the inside of the flask up and down at the surface of the solution with a glass rod. The scratching of the glass is thought to release microcrystals of glass, which serve as a template, or seeds, for crystal growth. If scratching the flask does not start the crystallization, a **seed crystal** may be added. A seed crystal is a small crystal of the original material set aside to provide a nucleus upon which other crystals can grow. Sometimes seed crystals can be obtained from the glass rod used for scratching, after the solvent has evaporated from it. Allowing a few drops of solution to

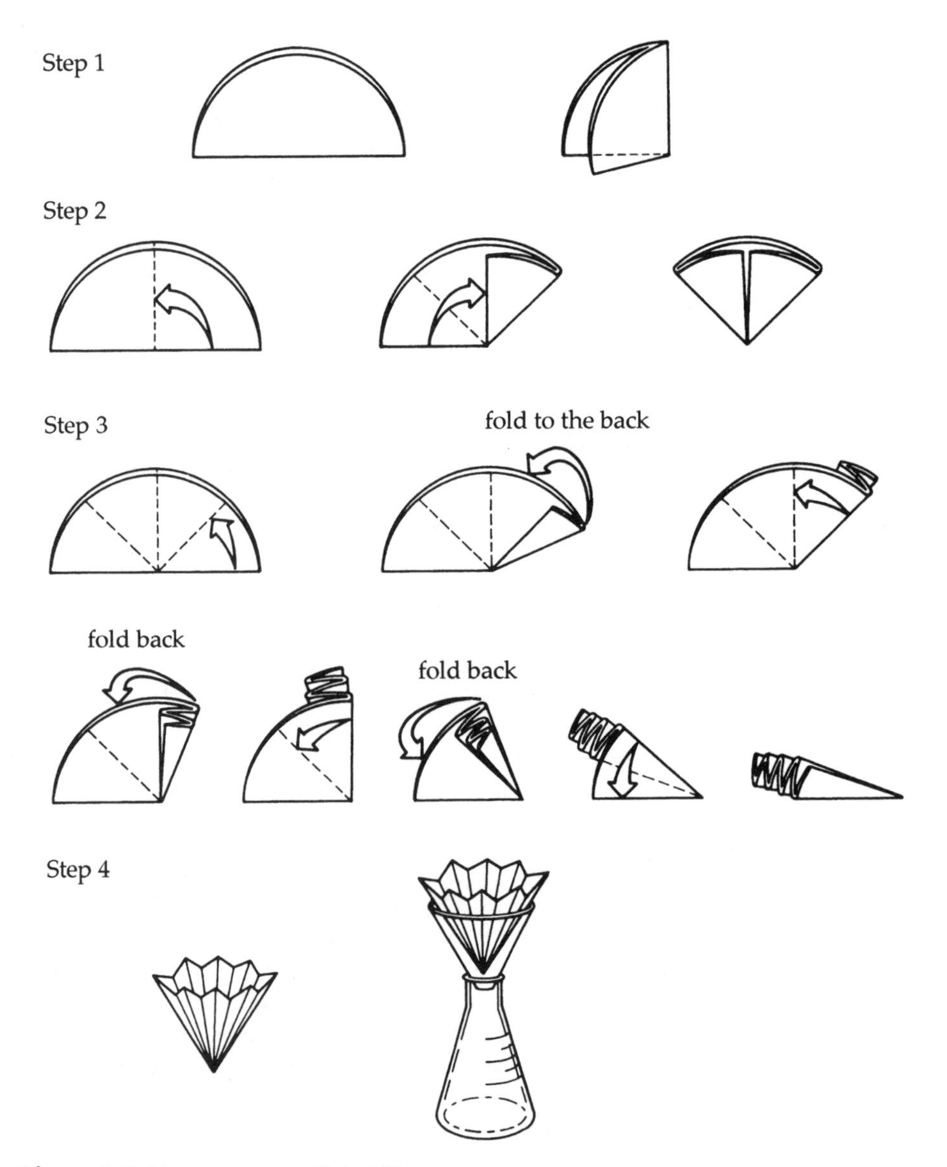

Figure 1.1 How to prepare fluted filter paper.

evaporate on a watch glass may also produce seed crystals. After addition of the seed crystal, set the flask aside to allow for crystallization to proceed.

If scratching and seeding do not produce crystals, your next question should be, "Did I use *too much* solvent?" If more than the minimum amount of solvent was used in the earlier steps, the excess must be boiled away (reduce the volume by about one-third) and the flask again set aside to crystallize.

Another problem encountered in crystallization is **oiling out**: Instead of crystals appearing, an oily liquid separates from solution. A compound may oil out if its melting point is lower than the boiling point of the solvent. A very impure compound may oil out because the impurities depress its melting point. The formation of an oil is not selective, as is crystallization; therefore, the oil (even if it solidifies) is probably not a pure compound.

Reheat a mixture that has oiled out in order to dissolve the oil (add more solvent if necessary). Then allow the solution to cool slowly, perhaps adding a seed crystal or scratching with a glass rod. If the substance has a low melting point, a lower-boiling solvent may be necessary. Alternatively, use more solvent and keep the temperature of the solvent below the melting point of the solute.

If these techniques do not prevent oiling out, allow the oil to solidify (a seed crystal or chilling may be necessary), filter the solid or decant (pour off) the solvent, and crystallize the solid using fresh or a different solvent. Enough impurities may have been removed in the attempted crystallization that the second one will proceed smoothly.

(4) Isolating the Crystals

Crystals are separated from their mother liquor by filtration. **Vacuum filtration** is the procedure of choice in all situations unless a very low-boiling crystallization solvent has been used. If a low-boiling solvent is used, the crystals are separated by gravity filtration.

Vacuum filtration apparatus. If water or a high-boiling organic solvent (such as the alcohols and most hydrocarbon solvents) has been used, then vacuum or suction filtration is used. Vacuum filtration has the advantage of being much faster than gravity filtration. It has the disadvantage of requiring more equipment. Figure 1.2 shows the physical setup required for vacuum filtration. The trap is necessary regardless of whether a water aspirator or a centralized vacuum system is used. The purpose of this trap is to (1) prevent any solution from being accidentally sucked into the vacuum line and (2) prevent any water from an aspirator from backing up into the filter flask. With a trap, this water will be caught before it contaminates the mother liquor.

Heavy-walled vacuum tubing must be used for vacuum connections, because ordinary tubing collapses when vacuum is applied. All flasks should be clamped to ring stands. The filter flask, especially, should be firmly clamped, because it usually becomes top-heavy when a Büchner funnel and vacuum line are connected to it.

Attach a Büchner funnel or Hirsch funnel to the filter flask with a rubber adapter or a one-holed rubber stopper (best) so that the connection will be airtight when vacuum is applied. Place a medium- or slow-speed filter paper (such as Whatman's No. 2, 5, or 6) on the perforated surface of the funnel. (A fast-speed, porous filter paper allows finely divided solids to pass through under vacuum.) The filter paper must lie flat and not curl up at the sides, yet it must cover all the holes. When the vacuum is applied, the filter paper is pulled snugly to the flat surface of the funnel by suction. To ensure no leakage around the edges, moisten the filter paper with the solvent before applying vacuum.

Water aspirator. Many laboratories are equipped with **water aspirators**, devices that attach to faucets and develop a vacuum through a side tube when water flows through the main tube. Place a large beaker in the sink under the water outlet to minimize splashing. Because aspirators are easily plugged, they should be checked before each use. Turn the water on *full force* and hold your finger on the

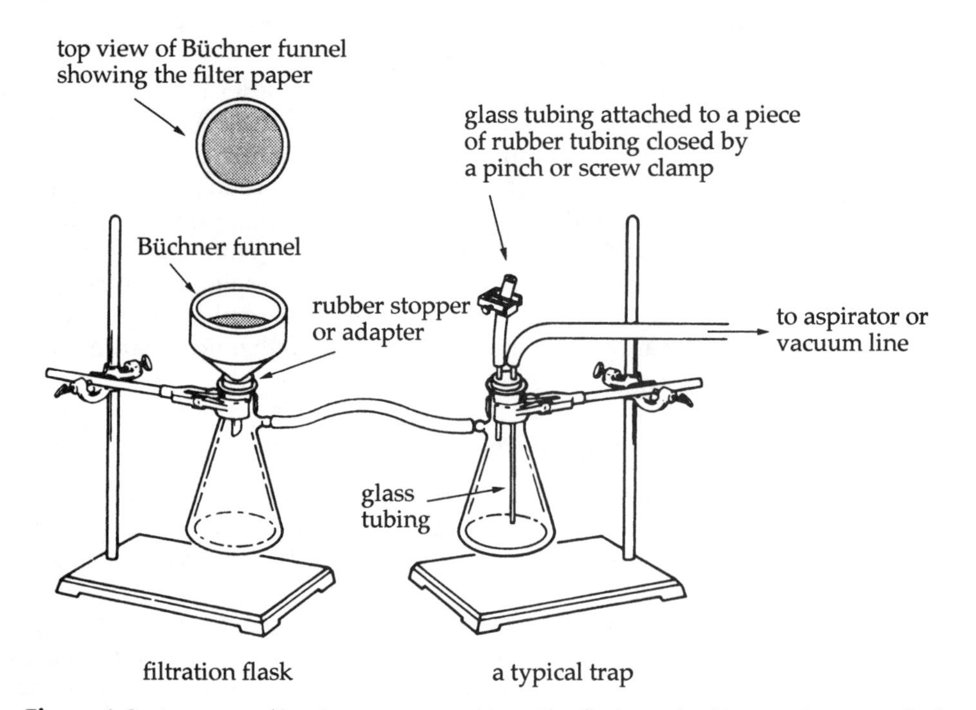

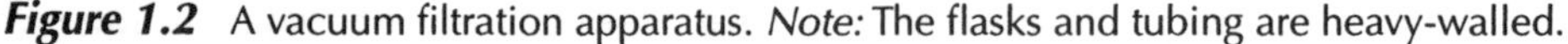

Figure 1.2 A vacuum filtration apparatus. *Note:* The flasks and tubing are heavy-walled.

vacuum hole to feel the suction before attaching the rubber tube of your filtration apparatus.

As previously mentioned, an aspirator can "back up." A slight decrease in water pressure can result in a greater vacuum in your filtration apparatus, causing it to suck water back into the apparatus. If you see water entering the trap, break the vacuum by opening the stopcock or pinch clamp on the trap, and then turn off the water.

The actual filtration. For nonvolatile solvents like water, apply the vacuum and pour the crystallization mixture into the Büchner funnel at such a rate that the bottom of the funnel is always covered with some solution. For high-boiling organic solvents, pour an initial portion of the mixture into the funnel; then apply the vacuum. In both cases, when the vacuum is applied, the mother liquor is literally sucked through the filter paper into the filter flask while the crystals remain on the filter paper. When the mother liquor ceases to flow from the funnel stem, release the vacuum by opening the stopcock on the trap. Then turn off the aspirator or vacuum line.

Washing. To wash the contaminating mother liquor from the crystals, transfer the crystal mass, or *filter cake,* to a small beaker, using a spatula to loosen, remove, and scrape the filter paper. Place fresh filter paper in the Büchner funnel, stir the crystals with a small amount of *chilled* solvent, and then immediately refilter. Small amounts of crystals may be washed right in the funnel on the original filter paper. This procedure is not recommended, because the wet filter paper may tear when you stir the wash solvent into the crystals and because this type of washing is not as thorough as a beaker washing.

Remove excess solvent from the crystals by putting a fresh piece of filter paper *on top of the crystals* still in the funnel and pressing this filter paper down firmly and all over with a cork. Keep the vacuum on during this pressing. When as much solvent has been pressed out of the filter cake as possible, leave the vacuum running for another minute or so. The air pulled through the filter cake will remove even more solvent. Then, open the trap clamp, turn off the vacuum, remove the Büchner funnel, and disconnect the filter flask assembly from the vacuum line. Using a spatula, pry the filter cake from the funnel for drying. The filter cake will often adhere to wet filter paper. Scrape the crystals from the paper only after it has dried.

Do not discard the mother liquor (in the filter flask), but place it in a corked Erlenmeyer flask until the completion of the experiment. The reason for saving the mother liquor is that it may still contain a substantial amount of the desired compound. Until you can determine a percent recovery of yield, you will not know if it is worthwhile to attempt to recover more material.

Gravity filtration. Gravity filtration is used if the crystallization solvent is low-boiling, for instance, diethyl ether or methylene chloride. Vacuum filtration should not be used for these solvents, because the solvent will evaporate during the filtration process and contaminate the crystals with the impurities that have just been removed.

Also called "simple filtration," gravity filtration employs a stemmed or stemless funnel and fluted filter paper. Select the size of filter paper that, when folded, will be a few millimeters below the rim of the funnel. Support the funnel in a ring or place it in the neck of an Erlenmeyer flask (use a paper clip to keep the funnel slightly away from the flask). Wet the filter paper with a few milliliters of cold solvent. Pour the mixture to be filtered through the funnel, in portions if necessary. Rinse the crystals on the filter paper with a small amount of chilled solvent, or use the beaker washing method described above.

(5) Drying the Crystals

The filter cake removed from the Büchner funnel or the gravity filtration filter paper still contains an appreciable amount of solvent. The crystals must be dried thoroughly before they can be weighed or before a melting point can be taken.

There are many methods of drying crystals. The simplest is *air-drying*, in which the crystals (with any lumps crushed) are spread out on a watch glass or large piece of filter paper and allowed to dry. Air-drying is sometimes slow, especially if water or some other high-boiling solvent was used. Unless the crystals are partially covered, they can collect dust. Another watch glass or a beaker, propped on corks to allow air to get to the crystals, may be used as a cover. For a melting-point determination, a few crystals may be removed from the mass and allowed to air-dry on a separate, uncovered watch glass. If a compound is hygroscopic (attracts water from the air), it cannot be air-dried.

In some laboratories, particularly in large student laboratories, drying chemicals on watch glasses in an open room that has a static atmosphere is discouraged for health reasons. Before air-drying your product, check the policy in force in your laboratory.

A **desiccator** may be used for drying a water-crystallized or hygroscopic compound. A desiccant (drying agent) such as anhydrous calcium chloride is placed in the bottom of the desiccator; the shelf is inserted; then a watch glass or beaker holding the crystals is placed on the shelf. When the cover is in place, the desiccant attracts water from the atmosphere in the desiccator as the water evaporates from the crystals. Some desiccators can be evacuated, thus speeding up the evaporation of any solvent.

(6) The Second Crop

The mother liquor from a crystallization may still contain in solution a large amount of the desired compound. In many cases, another batch of crystals, called the **second crop**, can be obtained from this mother liquor. To get the second crop, boil away one-third to one-half of the solvent from the mother liquor, and then allow it to cool and crystallize, as you did for the first crop of crystals. A seed crystal added to the cooled solution may help start the crystallization process. The second crop of crystals is rarely as pure as the first crop, because the impurities have been concentrated in the mother liquor. The purity of the second crop can be improved by recrystallizing these impure crystals with fresh solvent. Do not combine the first and second crops unless they are equal in purity.

The recovery in the second crop is usually reported separately from that in the first crop. In your notebook, you might write:

% recovery (1st crop) = 76%
% recovery (2nd crop) = 5.5%
total % recovery = 81.5%

When no more crops are to be isolated, discard the crystallization solvent in a designated waste container.

Safety Notes Summary: Crystallization

- Boiling chips should never be added to a hot solution.
- Toxic solvents should be heated only in a hood.
- A Bunsen burner should be used only for aqueous solutions—and only when no flammable solvents are being used in the vicinity.
- Some hot-plate surfaces are capable of igniting flammable solvents.
- The safest source for heating a solvent, especially a low-boiling solvent, is a steam bath. See the section on heating equipment in the Introduction (p. 15).

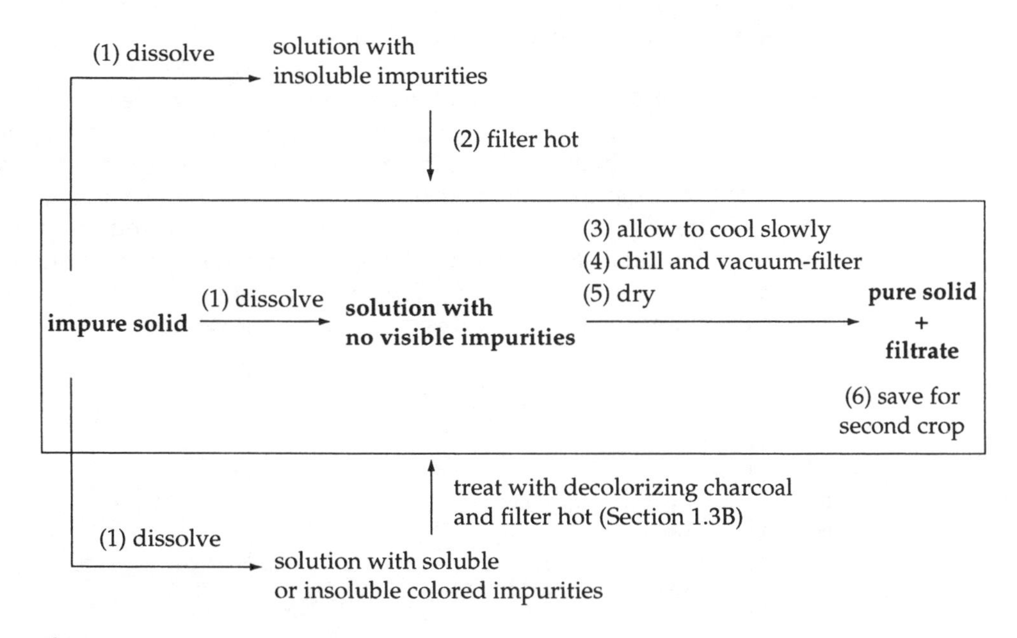

Figure 1.3 Steps in crystallization. (The numbers refer to the numbered steps in Section 1.2 of the text.)

1.3 Supplemental Procedures

A. Microscale and Semi-Microscale Procedures

When you need to recrystallize 10–100 mg of compound, the volume of solvent required is less than 5 mL. When using relatively small amounts of solvent, Erlenmeyer flasks are not suitable because they are too large. (As a rule of thumb, the flask should be about half-filled during a crystallization.) Different types of smaller containers work well in microscale crystallizations, from simple test tubes to specialized crystallization glassware such as Craig tubes. Whatever the type of glassware used, the basic technique of crystallization does not change: Dissolve the compound in the least amount of hot solvent, remove insoluble impurities if necessary, allow the solution to cool slowly, isolate the crystals from the mother liquor, and dry the crystals.

Test tube. Choose a heavy-walled test tube that will easily contain the required volume of solution. Place the compound to be crystallized in the test tube and add enough hot solvent to dissolve the compound. Keep the tube warm. If there are undissolved impurities, transfer the hot liquid to a clean test tube using a Pasteur pipet as illustrated in Figure 1.4, leaving the undissolved impurities behind. Allow the solution to cool slowly, as described in step 3, Section 1.2.

There are a couple ways to isolate the crystals from the mother liquor in the test tube. You can use vacuum filtration as described in Section 1.2, step 4. Choose a small Hirsch funnel rather than a Büchner funnel for microscale filtrations, as the sloping sides and small filtering area of a Hirsch funnel are more suited to small amounts of crystals. Use a small amount of chilled solvent to transfer crystals that remain in the test tube to the Hirsch funnel.

Another crystal isolation method employs a Pasteur pipet and bulb. First depress the bulb on the pipet, then carefully reach into the test tube, placing the tip of the pipet flush against the glass at the bottom of the test tube. Slowly and carefully release the bulb, drawing the mother liquid slowly into the pipet, leaving the crystals behind in the tube. Sometimes it helps to put a small cotton plug in the tip of the Pasteur pipet to prevent crystals from being sucked into the pipet. To rinse the crystals, add a small amount of chilled solvent, mix gently, and then remove the solvent with a pipet. Transfer the crystals to a watch glass to dry as usual, or dry the test tube and its contents in a desiccator or under vacuum.

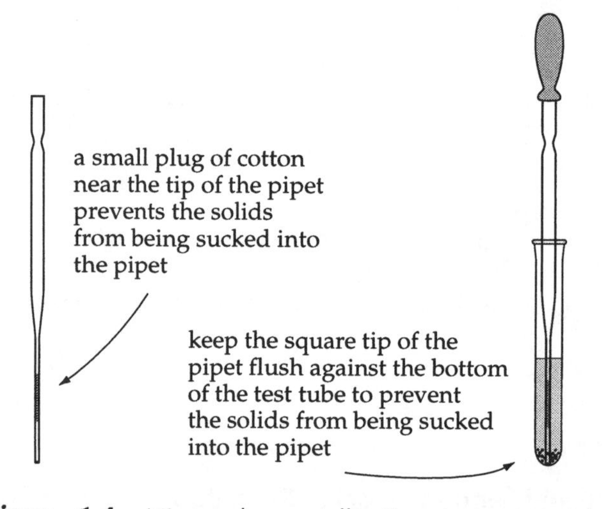
a small plug of cotton near the tip of the pipet prevents the solids from being sucked into the pipet
keep the square tip of the pipet flush against the bottom of the test tube to prevent the solids from being sucked into the pipet

Figure 1.4 Microscale crystallization in a test tube: removing hot or cold solvent from impurities or crystals.

Craig tube. A Craig tube consists of an outside tube and an inner plug (Figure 1.5). The compound to be crystallized is dissolved in hot solvent in a small vial or test tube and then transferred to the Craig tube. If there are undissolved impurities in the hot solvent, they are left behind during the transfer of hot solvent by using the technique illustrated in Figure 1.4. (The compound can be dissolved in the Craig tube itself, but only if there are no insoluble impurities.)

Place the Craig tube plug over the hot solution in the Craig tube. Support the assembly in a suitable container, such as an Erlenmeyer flask. Allow it to cool. When crystal formation is complete, place a centrifuge tube over the Craig tube and invert the whole assembly. Centrifuge the assembly to separate the mother liquor from the crystals. Disassemble the apparatus and carefully remove the crystals, which are now on the plug of the Craig tube apparatus.

Craig tubes have the advantage that they allow for very efficient separation of crystals from the mother liquid, as well as for fast drying of the crystals. Also, they lessen the number of transfers of solid material, thus increasing the yield of product. This is very important when working with the small amounts of material involved in microscale experiments.

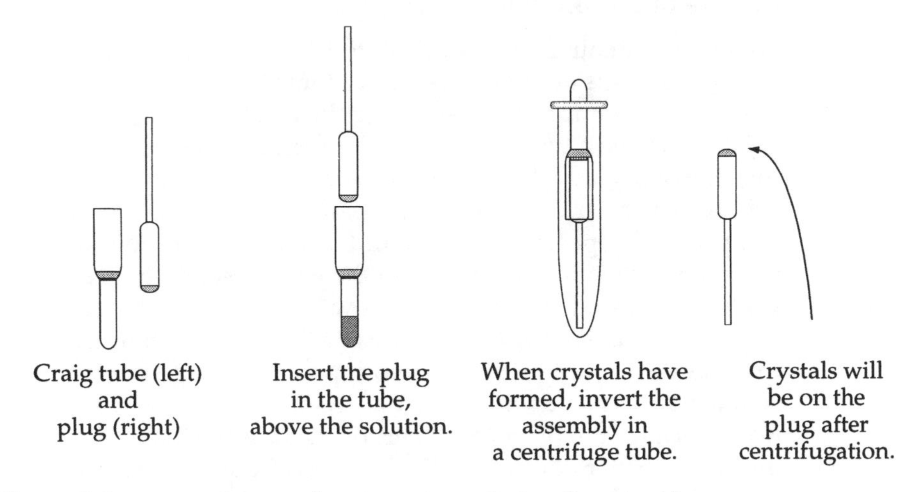

Craig tube (left)
and
plug (right)
Insert the plug
in the tube,
above the solution.
When crystals have
formed, invert the
assembly in
a centrifuge tube.
Crystals will
be on the
plug after
centrifugation.

Figure 1.5 Microscale crystallization using a Craig tube assembly.

B. Choosing a Solvent

If you do not know the appropriate solvent to use in crystallizing a compound, you must determine experimentally the solubility of the compound in various solvents. The following procedure can be used for this determination.

Weigh exactly 0.10 g of the compound into a small test tube. Pipet 1.0 mL of water into another test tube. Use these two test tubes as references for estimating 0.10 g of compound and 1.0 mL of solvent for the other test tubes.

Place an estimated 0.10 g of the compound into a series of small test tubes and add 1.0 mL of a different solvent to each. Try solvents of different polarities (Table 1.1). Stir each sample and determine the solubility of the compound in each solvent at room temperature. Record in your notebook whether the compound is insoluble (no apparent solid dissolved), slightly soluble (some of the solid dissolved), or soluble (no solid remains).

Place each test tube in a beaker of warm or boiling water, depending on the boiling point of the solvent. If some of the solvent should boil away, replace it with fresh solvent to maintain the volume at 1.0 mL.

Note the solubilities of the compound in the hot solvent. Some compounds may contain insoluble impurities even though the compound itself dissolves. Return the test tubes to a rack and allow them to cool to room temperature. Finally, place the test tubes in an ice bath. Record your observations.

From the preceding tests, choose the best solvent or solvent pair (Section 1.1); then weigh and crystallize the remainder of your sample, using about 10 mL of solvent per gram of solute. The compound in the test samples can be recovered by evaporating the test solvents in the fume hood and combining the solid material with the main sample before crystallization. (Evaporation of aqueous mixtures to recover material is usually not successful.)

C. Use of Decolorizing Charcoal

Frequently, small amounts of colored compounds and tarry (long-chain, or polymeric) materials are found as colored impurities in colorless organic compounds. These colored impurities can cause the crystallization solution and even the final crystals to have a tinge of color. Colored impurities can be removed with **decolorizing charcoal**, also called *activated charcoal* or *activated carbon*. Decolorizing charcoal, sold under the trade name Norit™, has a large surface area and adsorbs organic compounds, especially colored and polymeric compounds. Norit is added to the initial crystallization solution after the impure solid has been dissolved in warm solvent. Norit is sold either as a fine powder or as small pellets.

Norit powder. If the fine powder is used, the hot solution must be removed from the heat source before the Norit powder is added or the addition will cause the solution to bump and boil over the top of the flask. Only a small amount should be used, because the particles of carbon can adsorb the desired compound as well as impurities. After adding the charcoal, swirl the mixture a few times, and then carefully reheat. (Boiling solutions containing powdered decolorizing charcoal have a tendency to froth.) The Norit powder is then removed by hot filtration (see step 2, Filtering Insoluble Impurities, p. 26) before the solution is allowed to cool.

Pelletized Norit. If the Norit pellets are used, they are added to the hot solution and the solution is swirled for about 5 minutes—until the color disappears. During this process, the flask is kept hot. Then the pellets are "removed" by decanting the solution into a clean flask before the solution is allowed to cool.

Because the use of decolorizing charcoal always results in the loss of some of the compound being crystallized, decolorizing is carried out only when necessary and not as a routine procedure. If you are considering using decolorizing charcoal, first test an aliquot of the solution and judge the result.

D. Use of Filter Aids

Gelatinous precipitates, such as those of metal hydroxides and metal oxides, are difficult to remove from solutions of organic compounds because they clog the filter paper. Finely divided contaminants, such as decolorizing charcoal, are also difficult to remove because they may pass through filter paper. To circumvent these problems, use a filter aid (Filter-Cel®, Celite©). Filter aids are useful only in removing a solid contaminant; they cannot be used for filtering a desired solid.

Filter aids are diatomaceous earths (silica), which have large surface areas and are thus good adsorbents for contaminating solids. There are two techniques for using a filter aid: (1) Add the filter aid directly to the solution to be filtered, then heat or shake the mixture, and finally filter the mixture with vacuum. (2) Vacuum-filter a slurry of the filter aid and fresh crystallization solvent, and then filter the organic solution through the layer of filter aid resting on the filter paper. With either technique, use enough filter aid to cover the filter paper to a depth of 2–3 mm.

After filtering the solution, carefully wash the filter aid with fresh solvent to remove any adsorbed organic compound, refilter, and combine the filtrates.

Additional Resources

A list of suggested readings for the technique of crystallization is published on the Brooks/Cole Web site, as well as additional problems. Please visit www.brooks-cole.com.

Problems

1.1 Each of the following compounds, A–C, is equally soluble in the three solvents listed. In each case, which solvent would you choose? Give reasons for your answer. (More than one answer may be correct.)
(a) Compound A: benzene, acetone, or chloroform
(b) Compound B: chloroform, methylene chloride, or ethyl acetate
(c) Compound C: methanol, ethanol, or water

1.2 Which of the following solvents could not be used as solvent pairs for crystallization? Explain. (Hint: A solvent pair must be two miscible liquids; see Table 3.1, p. 51.)
(a) Hexanes and water
(b) Chloroform and diethyl ether
(c) Acetone and methanol

1.3 Suggest possible crystallization solvents for the following compounds.

(a) Naphthalene, [structure] (mp 80°)

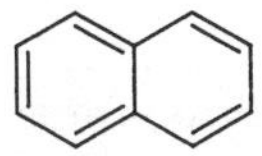

(b) Succinic acid, $HO_2C(CH_2)_2CO_2H$ (mp 188°)

(c) *p*-Iodophenol, I—[benzene ring]—OH (mp 94°)

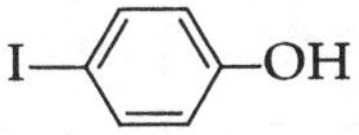

1.4 Give reasons for each of the following experimental techniques used in crystallization.
(a) A hot crystallization solution is not filtered unless absolutely necessary.
(b) An Erlenmeyer flask containing a hot solution is not tightly stoppered to prevent solvent loss during cooling.
(c) The suction of a vacuum filtration apparatus is broken before the vacuum is turned off.
(d) Vacuum filtration is avoided when crystals are isolated from a very volatile solvent.

1.5 A student was recrystallizing a compound. As the hot solution cooled to room temperature, no crystals appeared. The flask was then placed in an ice-water bath. Suddenly a large amount of solid material appeared in the flask. The student isolated a good yield of product; however, the product was contaminated with impurities. Explain.

1.6 A chemist crystallizes 17.5 g of a solid and isolates 10.2 g as the first crop and 3.2 g as the second crop.
(a) What is the percent recovery in the first crop?
(b) What is the total percent recovery?

1.7 A student crystallized a compound from benzene and observed only a few crystals when the solution cooled to room temperature. To increase the yield of crystals, the student chilled the mixture in an ice-water bath. The chilling greatly increased the quantity of solid material in the flask. Yet when the student filtered these crystals with vacuum, only a few crystals remained on the filter paper. Explain this student's observations.

1.8 The solubility of acetanilide in hot water (5.5 g/100 mL at 100°) is not very great, and its solubility in cold water (0.53 g/100 mL at 0°) is significant. What would be the maximum theoretical percent recovery (first crop only) from the crystallization of 5.0 g of acetanilide from 100 mL of water (assuming the solution is chilled to 0°)?

1.9 A 1.0 g sample of benzoic acid is contaminated with 0.05 g of salicylic acid. Solubilities in water of the two compounds are given in the following table.

Compound	*Solubility at 20° C (g/10 mL)*	*Solubility at 100°C (g/10 mL)*
benzoic acid	0.029	0.680
salicylic acid	0.22	6.67

(a) What volume of boiling water is needed to dissolve the 1.0 g of benzoic acid?
(b) How much benzoic acid will crystallize after cooling to 20°C?
(c) Will any salicylic acid crystals also form?
(d) Will the benzoic acid be pure?

1.10 Consider a sample of 1.5 g of benzoic acid contaminated with 1.5 g of salicylic acid (see Problem 1.9).
(a) What volume of boiling water is needed to dissolve the 1.5 g of benzoic acid?
(b) How much benzoic acid will crystallize after cooling to 20°C?
(c) Will any salicylic acid crystals also form?
(d) Will the benzoic acid be pure?

1.11 During a crystallization, while heating a solution of a compound to dissolve it in hot solvent, you boil it so long that a substantial amount of the solvent evaporates. What is likely to happen to some of the solute? What should you do if this occurs?

Melting Points

The **melting point** of a crystalline solid is the temperature at which the solid changes to a liquid at 1.0 atmosphere of pressure. The melting point is the same as the freezing point, the temperature at which the liquid becomes solid. Because liquids have a tendency to become supercooled (remain liquid below their freezing points), freezing-point determinations are only rarely performed in organic chemistry.

2.1 Characteristics of Melting Points

The melting point of a solid should be reported as a **melting range**. The barometric pressure, which has a negligible effect on the melting points at the usual atmospheric pressures, is ignored.

The melting point is determined by heating a small sample of the solid material slowly (at the rate of about 1° per minute). The temperature at which the first droplet of liquid is observed in the solid sample is the lower temperature of the melting range. The temperature at which the sample finally becomes a clear liquid throughout is the upper temperature of the melting range. Thus, a melting point might be reported, for example, as mp 103.5°–105°.

A. Effect of Impurities

A pure organic compound usually has a "sharp" melting point, which means that it melts within a range of 1.0° or less. A less pure compound exhibits a broader range, maybe 3° or even 10°–20°. For this reason, a melting point can often be used as a criterion of purity. A melting range of 2° or less indicates a compound pure enough for most purposes. However, a compound purified for spectroscopic analysis or for submission to an analytical laboratory for elemental analysis (determination of the relative weight percentages of the elements) should have a very sharp melting point.

An impure organic compound exhibits not only a broad melting range but also a *depressed* (lower) melting point from that of the pure compound. For example, a fairly pure sample of benzoic acid might melt at 121°–122°, but an impure sample might show a melting range of 115°–119°.

B. Using Melting Points to Identify Unknowns

Just comparing the melting point of an unknown with a literature value is insufficient evidence to identify an unknown. There are literally hundreds of different compounds that have the same melting point. However, even if two different compounds have the same melting point, if you mix the two together and take a melting point, the mixture will have a melting point that is broad and depressed. This is called a **mixed melting point**. Before the advent of modern spectroscopic techniques, mixed melting points were a powerful tool in compound identification.

For example, assume that you have a compound of unknown structure that melts at 121°–122°. Is the compound benzoic acid? To find the answer, you would mix the unknown with an authentic sample of benzoic acid (mp 122.4°) and then take a mixed melting point of the mixture. If the unknown is benzoic acid, the mixed melting point would remain 121°–122°, because the two mixed samples are the same compound. However, if the unknown is *not* benzoic acid, the mixed melting point would be depressed and show a wider range.

For absolute identification purposes, additional data besides the mixed melting point are desirable.

C. Melting-Point Diagrams

The melting-point diagram in Figure 2.1 shows the typical melting behavior of a series of mixtures of two organic compounds, *o*-dinitrobenzene and *m*-dinitrobenzene.

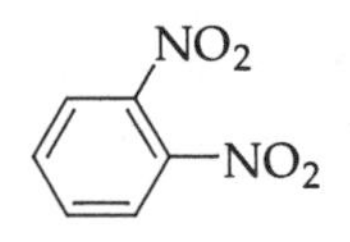

o-dinitrobenzene

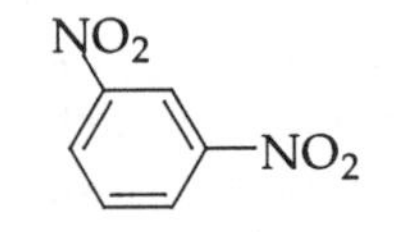

m-dinitrobenzene

A graph for mixtures of two other compounds would probably be similar, although different types of melting behavior are occasionally observed (Section 2.1D). The graph shows the upper limit of the melting range (the temperature at which the mixture becomes completely liquid) and the lower limit of the melting range (the temperature at which the mixture first begins to melt). The graph also shows that pure *o*-dinitrobenzene melts at 118.5°, pure *m*-dinitrobenzene melts at 90.0°, and a 50:50 (molar ratio) mixture of the two compounds melts at about 80°.

The lower point on the graph (63°) is called the **eutectic point**, or **eutectic temperature**, and it is the lowest melting point of any mixture of these two compounds. The percent composition of the mixture that melts at the eutectic point is called the **eutectic mixture**. The values for the eutectic point and the percent composition of a eutectic mixture depend on which compounds are being studied. For the diagram in Figure 2.1, the eutectic mixture consists of 63% *m*-dinitrobenzene and 37% *o*-dinitrobenzene. Binary mixtures of other compounds will have different values for their eutectic temperature and mixture. The eutectic point for any binary mixture is the temperature at which both components melt simultaneously and is, consequently, a sharp melting point rather than the broad melting range usually observed for mixtures.

A eutectic point is exhibited only by an intimate and uniform mixture of the correct composition. For all practical purposes in the organic laboratory, virtually all mixtures of two different compounds exhibit a broad melting range.

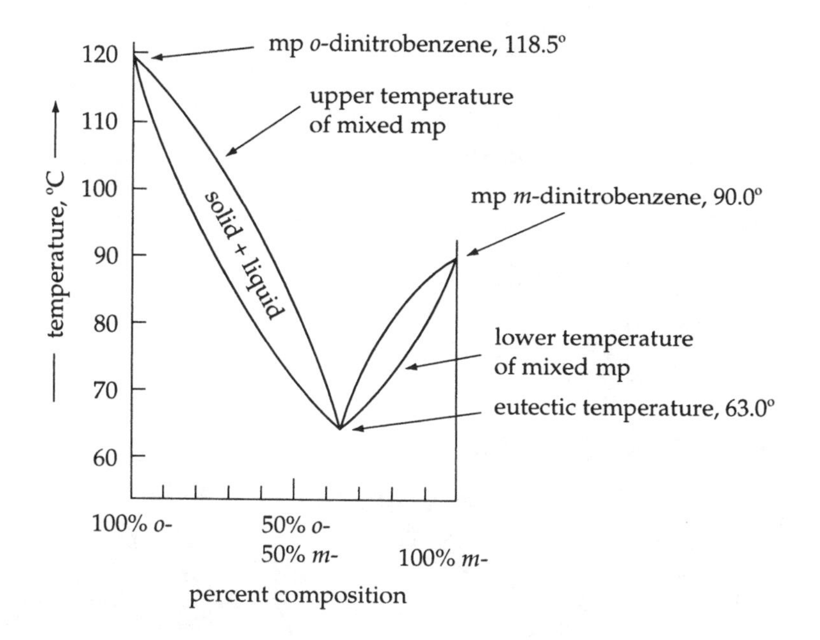

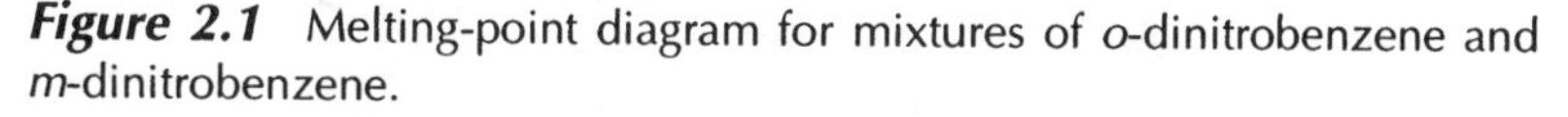

Figure 2.1 Melting-point diagram for mixtures of *o*-dinitrobenzene and *m*-dinitrobenzene.

D. Other Melting Behavior

Decomposition. All organic compounds decompose when heated to sufficiently high temperatures. Many organic compounds decompose at or below their melting points. Some of these compounds exhibit sharp melting points with evidence of decomposition, such as darkening. Others, even pure compounds, may exhibit a broad melting–decomposition range.

Polymorphs. Some compounds exhibit polymorphism. Polymorphs are different crystalline forms of the same substance. Variations in the intermolecular attractions of the different crystalline forms can cause polymorphs to have different, but sharp, melting points. Which crystalline form is encountered depends on such factors as the temperature at which crystallization occurred, the rate of crystallization, and the solvent.

When more than one melting point for a pure organic compound is reported in the literature, it generally means that the compound forms a polymorph.

Hydrates. Some compounds can crystallize with water or other solvent molecules incorporated in their crystal lattice in a definite proportion by weight. In the case of water, these molecules are called "water of hydration" and the combination of compound and water is called a "hydrate."

A hydrate melts at a different temperature than does the anhydrous form of the compound. For example, the hydrate of oxalic acid melts at 101.5°, but anhydrous oxalic acid exists in two polymorphic forms, one melting at 182° and the other at 189.5°.

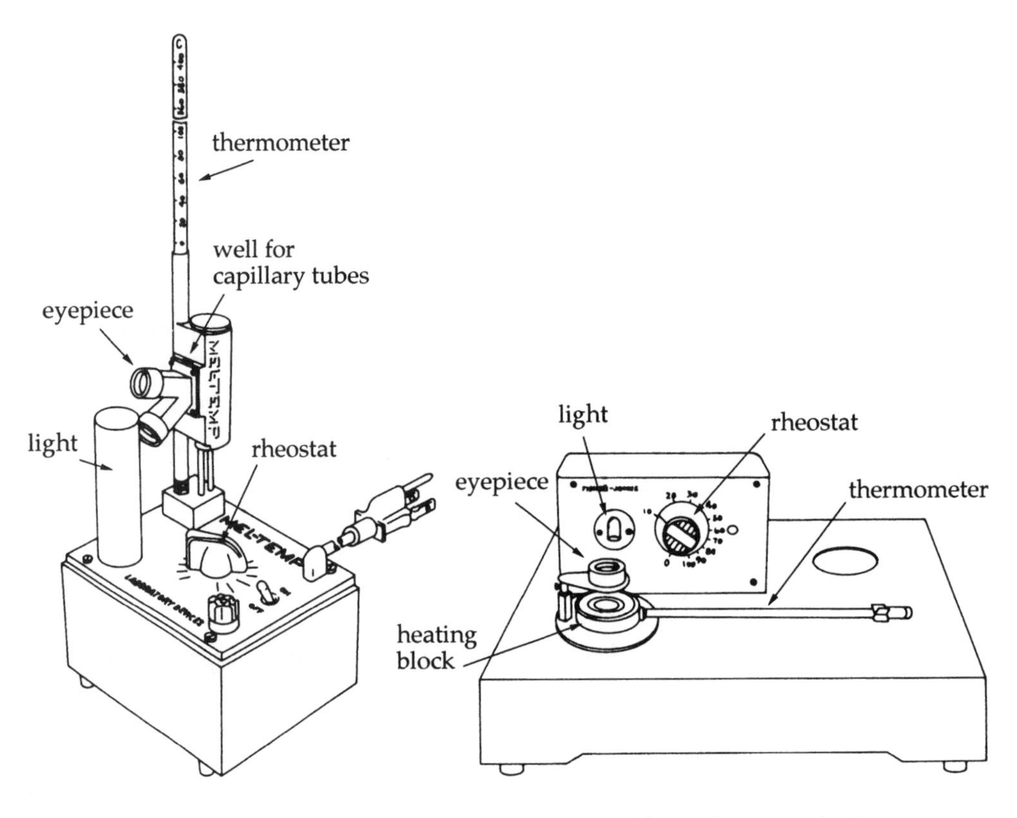

Figure 2.2 Two types of melting-point devices. Melting-point capillaries containing samples are inserted into the capillary well in the capillary melting-point device. Crystals are placed between microscope cover glasses and put on the heating block in the melting-point stage device.

2.2 Melting-Point Apparatus

Many types of electrically heated melting-point devices are on the market today; two are shown in Figure 2.2. With one type of device, a glass capillary tube containing the sample is inserted into a heating block, and the melting of the compound is observed through a magnifying eyepiece. Most instruments will accept several capillary tubes simultaneously, so that the melting points of several samples can be determined under identical conditions. Simultaneous melting-point determinations are especially useful for mixed melting points, because the melting behavior of the pure compound and of one or more mixtures can be compared.

A *melting-point stage* is an apparatus in which the sample (as little as a single crystal) is placed between two microscope cover glasses instead of in a capillary. The cover glasses are then placed on an electrically heated block (the stage), and the melting behavior is observed through a magnifying glass. A melting point determined on a stage may be slightly higher than the capillary melting point, because the sample is not actually submerged in the heat source. In the following instructions, we assume you will be taking capillary melting points.

A rheostat controls the rate of heating of an electrical melting-point apparatus. The higher the setting on the rheostat dial, the faster the block will heat. Manufacturers measure the rate of heating at different dial settings, graph the values, and enclose this data with each instrument shipped. They also include recommended dial settings for different melting temperatures. (See Figure 2.3.) This is important information, because when taking a melting point, you need to have the block heating at a rate of about 1° per minute in the vicinity of the melting point.

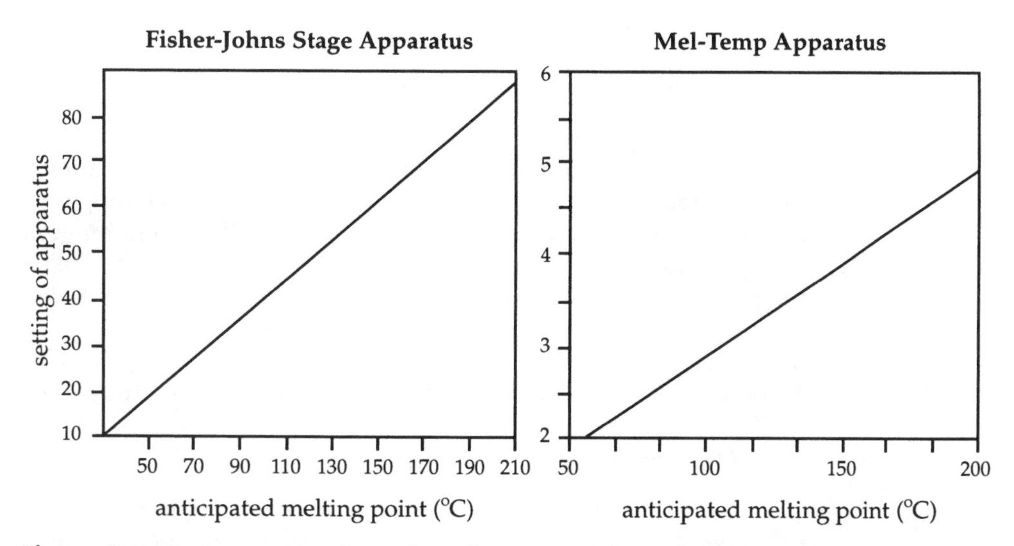

Figure 2.3 Proper settings for Fisher-Johns stage (left) and Mel-Temp capillary (right) melting-point devices.

2.3 Steps in Determining a Melting Point

(1) Preparation of the Sample

Pulverize the sample for a melting-point determination by placing 0.1–0.2 g of dry crystals in a watch glass and crushing them with a metal spatula or the end of a test tube. If the sample is to be used for a mixed melting point, grind a 50:50 mixture of the two compounds (approximated, not necessarily weighed) thoroughly with a mortar and pestle to ensure a homogeneous and intimate mixture. Alternatively, dissolve the mixture of the two compounds in a suitable solvent on a watch glass, allow the solvent to evaporate, and then pulverize the residue.

(2) Loading the Capillary

Mound up the pulverized sample and press the open end of the capillary into the sample against the surface of the watch glass or mortar. A small plug of sample will be pushed into the open end of the capillary tube. The ideal amount of sample is a plug about 1 mm in length. Tapping the sealed end of the capillary on the surface of the laboratory bench may knock the sample down to the desired position at the sealed end of the capillary. (CAUTION: Capillaries are fragile.) A safer and more efficient technique for driving the plug of sample to the sealed end of the capillary tube is to drop the tube (sealed end down) through 2–3 feet of ordinary glass tubing vertically onto the benchtop. The impact of the capillary with the hard surface knocks the sample down. It may be necessary to drop the capillary two or three times.

It is important that the height of the sample in the capillary be only 1–2 mm and that the sample be packed firmly. A larger sample takes longer to melt and may exhibit an erroneously large melting range. If the sample is loosely packed, it is difficult to determine the start of the melt.

(3) Preliminary Melting Point

If the approximate melting point of a sample is not known, it saves time to take a preliminary melting point with a second capillary tube. The approximate melting point of the sample is determined by rapidly heating the capillary with the melting-point apparatus (about 10° per minute). This preliminary melting point will be on the low side of the true melting point, but it will give you an idea of the rate of temperature increase you should use to determine the exact melting point.

You need not determine a preliminary melting point if you know the name of the compound and can find its melting point in a reference book or journal. Refer to Technique 17, p. 198, for sources of physical data of organic compounds.

(4) Taking the Melting Point

Insert the capillary tube containing the sample into the melting-point apparatus, along with the thermometer (if necessary). Heat the apparatus rapidly to a temperature about 10° below the expected melting point, and then slow the rate of temperature increase to about 1° per minute. If the temperature is increased too rapidly at the melting point, the thermometer, sample, and block will not be at thermal equilibrium, and erroneous readings will result.

As we have mentioned, the initial, or lower, end of the melting range is the temperature at which the first drop of liquid is noted in the solid sample. The final, or upper, end of the range is the temperature at which the sample becomes a clear liquid containing no solid material. The determination of the final value of the melting range usually presents no problems. The initial value, however, may require judgment. Many organic compounds undergo changes in crystal structure prior to melting, and these phase changes may be mistaken for the first sign of melting. Sample sag, shrinking, changes in texture, and the appearance of droplets *outside the bulk of the sample* are not the start of the melt. The initial temperature of the melting range is taken as the first appearance of liquid *within the bulk of the sample.*

Another problem that may arise is decomposition. In a simple case, a sample may change color, effervesce, or otherwise change in appearance at its melting point. If the melting point is reasonably sharp, this type of behavior does not affect the value of the melting point as a criterion of purity or as an identification tool, but the decomposition should be reported along with the melting point. For example:

"mp 150.3°–151.5°d" or
"150.3°–151.5° dec" or
"150.3°–151.5° with darkening"

If a sample decomposes over a large temperature range, the melting point cannot be used for identification purposes. The decomposition range should be reported, even if only approximate temperatures can be given. For example:

"dec 127°– ~150°"

Used capillary tubes cannot be cleaned for reuse. Follow the procedure in your laboratory for proper disposal of used melting-point capillary tubes.

Safety Notes

- Capillaries and melting-point slides are fragile. Handle them with care to avoid cutting yourself.
- Melting-point devices get very hot at the melting stages. Avoid touching the hot surfaces.
- Wear eye protection while taking melting points because some compounds produce irritating vapors as they melt.

2.4 Supplemental Procedures

A. Melting Point of a Volatile Solid

Some solids (such as camphor or menthol) vaporize before melting when they are heated. To determine the melting point of a volatile solid, the open end of the melting-point capillary must be sealed before the melting point can be taken.

To seal a melting-point capillary, first load the capillary in the usual way. Then, holding the capillary tube horizontally, twirl or roll the capillary with your fingers, having about a millimeter of the open end in the base of a flame. The edges of the open end will melt and collapse. The twirling or rolling of the tube back and forth ensures that the glass will collapse inward and seal the tube. After sealing the tube, inspect the closed end with a magnifying glass to be sure that it does not contain a pinhole.

B. Thermometer Calibration

No thermometer is accurate at every temperature reading; therefore, a thermometer to be used for melting-point determinations in a research laboratory (and sometimes in the student laboratory) should be calibrated. Calibration is accomplished by recording the melting points of five or six very pure compounds, chosen to melt at a variety of temperatures. Table 2.1 lists some suggested compounds. From these melting points, a graph similar to the one in Figure 2.4 is constructed. This graph shows the *correction factor* versus the *observed temperature* (using the upper value of

Table 2.1 Suggested melting-point standards for calibration of thermometers.

Compound	*Melting point (°C)*
naphthalene	80.6
1,3-dinitrobenzene	90.0
acetanilide	114.3
benzoic acid	122.4
benzamide	130
urea	135
sulfanilamide	166
p-toluic acid	182
succinic acid	188
3,5-dinitrobenzoic acid	205
anthracene	216.3
p-nitrobenzoic acid	242

the melting range). For example, pure benzoic acid melts at 122.4°. If your thermometer records 120.9°, your correction factor at approximately 120° would be +1.5°. Any time you record a melting point near 120°, you would add 1.5° to the observed temperature. (Corrected melting point = observed melting point + 1.5°.) A corrected melting point is reported as follows: mp 121.2°–121.5° (cor).

Melting points taken with a thermometer calibrated as described above need no additional correction factor for the exposed thermometer stem. If you use a thermometer that has been factory-calibrated while completely submerged, you may want to correct the melting points with *stem correction factors*. Calculations of stem correction factors are discussed in Table D of the *CRC Handbook of Chemistry and Physics* (Boca Raton, FL: CRC Press, Inc).

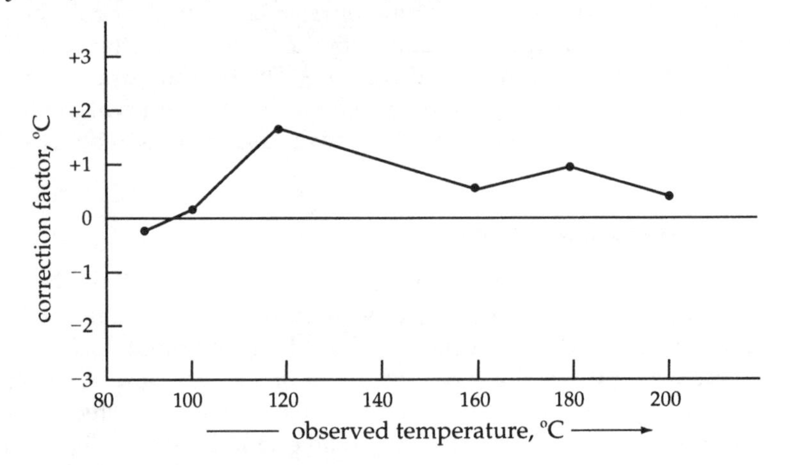

Figure 2.4 A typical thermometer calibration graph.

Additional Resources

A list of suggested readings pertinent to melting point determination, discussions of racemates and liquid crystals, and additional problems are published on the Brooks/Cole Web site. Please visit www.brookscole.com.

Problems

2.1 Look up the melting points of the following compounds in a suitable handbook, Internet site, or online source. (See Technique 17, p. 198.) Include the name of your source with your answer.
(a) Quinine
(b) *o*-Iodophenol

2.2 What setting should you have on the Mel-Temp apparatus when you are in the vicinity of the melting point for each of the compounds in Problem 2.1? (Refer to Figure 2.3.)

2.3 What setting should you have on the Fisher-Johns apparatus when you are in the vicinity of the melting point for each of the compounds in Problem 2.1? (Refer to Figure 2.3.)

2.4 Using the following table, draw a melting-point diagram.

Percent composition	*Melting range (°C)*
100% A	125°–126°
75% A–25% B	115°–120°
65% A–35% B	128°–31°
50% A–50% B	135°–140°
100% B	151°–152°

(a) Estimate the eutectic temperature and composition.
(b) If you wished to obtain a more accurate graph for this melting-point curve, suggest the best mixtures for additional mixed melting points.

2.5 Why is it *not* good practice to mix two components by simply stirring them together and then sampling them for a mixed melting point?

2.6 You think that you have isolated aspirin (acetylsalicylic acid) in the lab. It melts at 75°–77°C. Since you don't totally trust your own laboratory techniques, you want to prove to yourself that you have aspirin. Using only melting-point techniques, explain how you can prove that you actually have aspirin. (Assume the stockroom is able to supply you with any compound you need.)

2.7 The following table lists the observed melting points for several compounds isolated in the student laboratory.

Compound	*Observed melting point (°C)*
naphthalene	79°–80°
benzophenone	45°–47°
p-anisic acid	178°–182°
salicylic acid acetate	135°
3-chlorobenzoic acid	157°–158°
sulfanilamide	165°–166°
ferrocene	157.5°–161.5°

(a) Look up the melting point for each compound in a suitable resource.
(b) Which compounds are pure? Which are impure?

2.8 Suppose that you are taking a melting point and the compound disappears: What happened? Why did it do this? What should you do?

2.9 You and your lab partner take melting points of the same sample. You observe a melting point of 101°–107°C, while your partner observes a value of 110°–112°C. Explain how two different melting point values can be observed for the exact same sample.

2.10 Using Table 2.1, construct a thermometer calibration graph from the following observed melting points.

Acetanilide, mp 113°
Benzamide, mp 130°
Succinic acid, mp 188°
p-Nitrobenzoic acid, mp 245°

Extraction

Extraction is the separation of a substance from one phase by another phase. The term is usually used to describe removal of a desired compound from a solid or liquid mixture by a solvent. In a coffee pot, caffeine and other compounds are extracted from the ground coffee beans by hot water. Vanilla extract is made by extracting the compound vanillin from vanilla beans.

In the laboratory, several types of extraction techniques have been developed. The most common of these is *liquid–liquid extraction*, or simply "extraction." Extraction is often used as one of the steps in isolating a product of an organic reaction. After an organic reaction has been carried out, the reaction mixture usually consists of the reaction solvent and inorganic compounds, as well as organic products and by-products. In most cases, water is added to the reaction mixture to dissolve the inorganic compound. The organic compounds are then separated from the aqueous mixture by extraction with an organic solvent that is immiscible with water. The organic compounds dissolve in the extraction solvent, while the inorganic impurities remain dissolved in the water.

The most commonly used device to separate the two immiscible solutions in an extraction procedure is the **separatory funnel**. Typically, the aqueous mixture to be extracted is poured into the funnel first, and then the appropriate extraction solvent is added. The mixture is shaken to mix the extraction solvent and the aqueous mixture, and then it is set aside for a minute or two until the aqueous and organic layers have separated. The stopcock at the bottom of the separatory funnel allows the bottom layer to be drained into a flask and allows the separation of the two layers, as shown in Figure 3.1. The result (ideally) is two separate solutions: an organic solution (organic compounds dissolved in the organic extraction solvent) and an inorganic solution (inorganic compounds dissolved in water). Unfortunately, often the water layer still contains some dissolved organic material. For this reason, the water layer is usually extracted one or two times with fresh solvent to remove more of the organic compound. In addition, some of the water layer is usually retained on the sides of the separatory funnel and contaminates the organic layer. This water contaminant is removed in a separate step later in the workup or isolation procedure (see Technique 4).

After one or more extractions and separations, the combined organic solutions are usually extracted with small amounts of fresh water to remove traces of inorganic acids, bases, or salts; treated with a solid drying agent to remove traces of water; and then filtered to remove the hydrated drying agent. Finally, the solvent is evaporated or distilled. The organic product can then be purified by a technique such as crystallization or distillation.

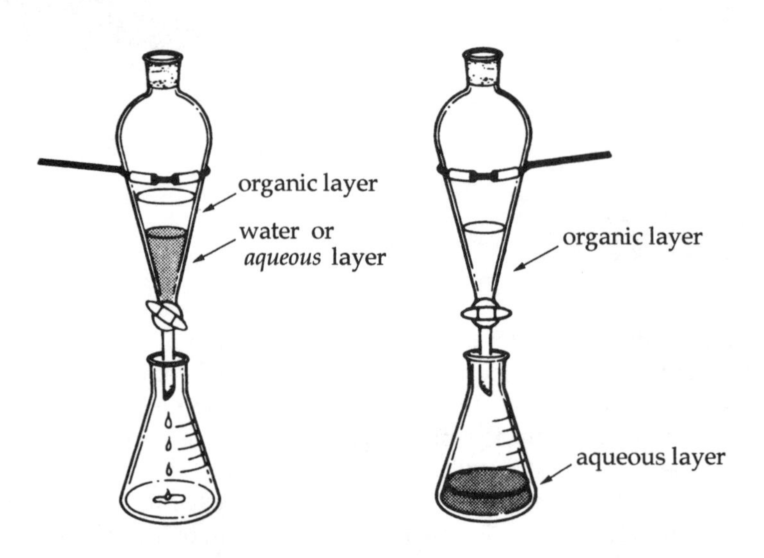

Figure 3.1 Two immiscible solutions can be separated with a separatory funnel. (The organic layer may be the upper or lower layer, depending on the relative densities of the two solutions.)

3.1 Immiscible Liquids

For a successful extraction of a compound from one liquid to another, the two liquids must be **immiscible**: not soluble in each other. Water is immiscible with most organic solvents: butanol, chloroform, cyclohexane, methylene chloride, ethyl acetate, hexanes, toluene, diethyl ether, and pentane. The miscibility properties of common extraction solvents are summarized in Table 3.1. Note that water is miscible with methanol, ethanol, isopropyl alcohol, and acetone. Therefore, these solvents cannot be used to extract a compound from an aqueous solution. Acetone and ethanol are miscible with every solvent listed in the chart; therefore, they are not used as extraction solvents.

The designation "immiscible" is somewhat subjective. Some solvent pairs designated immiscible in Table 3.1 are miscible to a small degree, as shown in Table 3.2. Note that water is somewhat soluble in diethyl ether and in ethyl acetate. Although these solvents can be used in an extraction of a compound from water, a small amount of the solvent and therefore a small amount of the compound being extracted will remain in the water, and water will remain in the organic solvent, which then must be dried (see Technique 4).

Table 3.1 Miscibility properties of some common solvents.*

	acetone	butanol	chloroform	cyclohexane	methylene chloride	ethanol	ethyl acetate	diethyl ether	hexanes	methanol	pentane	isopropyl alcohol	toluene	water
acetone														
butanol														
chloroform														
cyclohexane														
methylene chloride														
ethanol														
ethyl acetate														
diethyl ether														
hexanes														
methanol														
pentane														
isopropyl alcohol														
toluene														
water														

* Shading in a square indicates immiscibility.

Table 3.2 Solubility properties of some common solvents and water.

Solvent	*Solubility of water in solvent (g/100 mL solution)*	*Solubility of solvent in water (g/100 mL solution)*
hexanes	0.007	0.001
methylene chloride	0.32	1.6
diethyl ether	1.24	5.6
ethyl acetate	2.92	8.20

3.2 Distribution Coefficients

When a compound is shaken in a separatory funnel with two immiscible solvents, such as water and diethyl ether, the compound distributes itself between the two solvents. Some of the compound dissolves in the water and some in the ether. How much of the compound or solute dissolves in each phase depends on the solubility of the solute in each solvent. The ratio of the concentrations of the solute in each solvent at a particular temperature is a constant called the **distribution coefficient** or **partition coefficient (K)**.

$$K = \frac{\text{concentration in solvent}_2}{\text{concentration in solvent}_1}$$

where solvent_1 and solvent_2 are immiscible liquids

By convention, solvent_2 is the solvent in which the compound (solute) of interest is more soluble. In calculations of distribution coefficients, we assume that the solute neither ionizes in nor reacts with either solvent. Because a ratio is involved, the concentrations may be in any units, as long as the two concentrations are the *same* units.

To a rough approximation, the ratio of concentrations of the above equation is the same as the ratio of the *solubilities* of the compound in the two solvents, measured independently. For example, consider a compound, which we will call compound A, that is soluble in diethyl ether to the extent of 20 g/100 mL at 20°C and soluble in water to the extent of 5.0 g/100 mL at the same temperature. We can approximate the distribution coefficient of compound A in diethyl ether and water to be 4.0, where diethyl ether is solvent_2 in the following equation:

$$K = \frac{\text{concentration in solvent}_2}{\text{concentration in solvent}_1}$$

$$= \frac{20\ \text{g}/100\ \text{mL}}{5.0\ \text{g}/100\ \text{mL}}$$

$$= 4.0$$

Given the distribution coefficient for a particular system, we can calculate how much compound will be extracted. Supposed that we have a solution containing 5.0 g of compound A in 100 mL of water. If we shake this solution with 100 mL of diethyl ether, how much A will be extracted by the ether? For this calculation, we will assume that the ether–water distribution coefficient is 4.0.

To solve this problem, we will let x be equal to the number of grams of compound A in the diethyl ether after the liquids have been shaken. Since we started with 5.0 g of A in the water, the number of grams remaining in the water is $5.0 - x$. The respective concentrations are the number of grams in each layer divided by each volume.

$$\text{concentration of A in diethyl ether} = \frac{x\ \text{g}}{100\ \text{mL}}$$

$$\text{concentration of A in water} = \frac{(5.0 - x)\text{g}}{100\ \text{mL}}$$

By substitution,

$$K = \frac{\text{concentration in diethyl ether}}{\text{concentration in water}}$$

$$4.0 = \frac{x\ \text{g}/100\ \text{mL}}{(5.0 - x)\text{g}/100\ \text{mL}}$$

Solving,

$$4.0 = \frac{x}{5.0 - x}$$

$$20 - 4x = x$$

$$5.0x = 20$$

$$x = 4.0 \text{ g in 100 mL of diethyl ether}$$

From this calculation, we see that we will extract only 80% of A from the water solution. Therefore, 20%, or 1.0 g, of A will remain in the water layer.

Because a solute distributes itself between two solvents, a single extraction may not be very efficient. Considerable amounts of material may remain behind in the original solvent. Is it possible to make liquid–liquid extraction more efficient? Using the same amount of diethyl ether, can we extract more than 4.0 g of A from the original water solution? Yes, we can. Let us divide the 100 mL of ether into three portions of approximately 33 mL each. Then, let us extract the water solution three separate times, using 33 mL of fresh diethyl ether for each extraction. If you carry out the calculation for each extraction as before (allowing for the different volumes of solvent), you will find that a total of 4.5 g of A can be extracted.

Our conclusion is that it is a more efficient use of solvent to perform three small extractions than one large one. A greater number of small extractions would remove an even greater quantity of the solute from water. If compound A were valuable, we would perform the extra extractions; otherwise, we would not bother.

Most organic compounds have distribution coefficients between organic solvents and water greater than 4. Therefore, a double or triple extraction generally removes almost all of the organic compound from the water. However, for water-soluble compounds, where K may be less than 1, we can calculate that only a small amount of the compound will be extracted. We predict that the extraction will fail. In order to extract the compound, we will have to use continuous liquid–liquid extraction (Section 3.7A) or salting out (Section 3.6C).

3.3 Extraction Solvents

The preceding discussion has provided some clues for the choice of an extraction solvent. The extraction solvent must be immiscible with the first solvent, which is generally water. The compound to be extracted should be soluble in the extraction solvent and, of course, not undergo reaction with it. (Technique 1, p. 24, contains a brief discussion of solubilities.) Major impurities should not be soluble in the extraction solvent. In addition, the extraction solvent should be sufficiently volatile that it can be removed by distillation from the extracted material later in the workup procedure. It is also preferable that the solvent be nontoxic and nonflammable. Unfortunately, many organic solvents do not meet these last two criteria.

Diethyl ether is both volatile (boiling point only a few degrees above room temperature) and flammable. Benzene is toxic and flammable. Halogenated hydrocarbons are not all flammable, but most are toxic. When using a solvent for extraction, always proceed with caution. Table 3.3 lists a few extraction solvents, their densities, and their potential hazards.

Table 3.3 Some common extraction solvents.

Name	*Formula*	*Density (g/mL)**	*bp (°C)*	*Comments*
lighter than water:				
diethyl ether	$(CH_3CH_2)_2O$	0.7	35	highly flammable
petroleum ether	C_nH_{2n+2}	~0.7	~30–60	flammable
hexanes	C_6H_{14}	~0.7	67–69	flammable
benzene	C_6H_6	0.9	80	flammable, toxic, carcinogenic
toluene	$C_6H_5CH_3$	0.9	111	flammable
heavier than water:				
methylene chloride†	CH_2Cl_2	1.3	40	toxic
chloroform	$CHCl_3$	1.5	61	toxic
carbon tetrachloride	CCl_4	1.6	77	toxic, carcinogenic

* The density of water is 1.0 g/mL; that of saturated aqueous sodium chloride solution is 1.2 g/mL.

† Also known as dichloromethane.

Note in Table 3.3 that the chlorinated hydrocarbon solvents are more dense than water; these solvents sink to the bottom in a separatory funnel containing water. The other solvents listed usually float on water. An exception would be a solvent containing a high concentration of a dense solute, which can increase the density of the organic layer so that it becomes heavier than water. Prematurely discarding the wrong solution is a common error. For this reason, it is generally wise to test the two layers if there is any question as to which is the organic layer and which is the aqueous layer. A simple test is to add a few drops of each layer to a small amount of water in a pair of test tubes. The layer that is immiscible with water is the organic layer. Also, it is generally wise to save all layers until an experiment is complete.

3.4 Chemically Active Extraction

The type of extraction procedure that we have been discussing can be considered a "passive" process—the extraction of a compound by virtue of its distribution between a pair of solvents. A less common, but very powerful, extraction technique is **chemically active extraction**. In this type of extraction, a compound is altered chemically to change its distribution between a pair of solvents. The most common method of changing the chemical structure is by using an acid–base reaction.

To illustrate how chemically active extraction works, let us consider a specific example. Assume we have a mixture of two compounds, a hydrocarbon and a carboxylic acid. Further assume that the hydrocarbon and the carboxylic acid are both soluble in diethyl ether and insoluble in water. These two compounds cannot be separated from one another by passive extraction. For example, the following mixture is soluble in ether and insoluble in water:

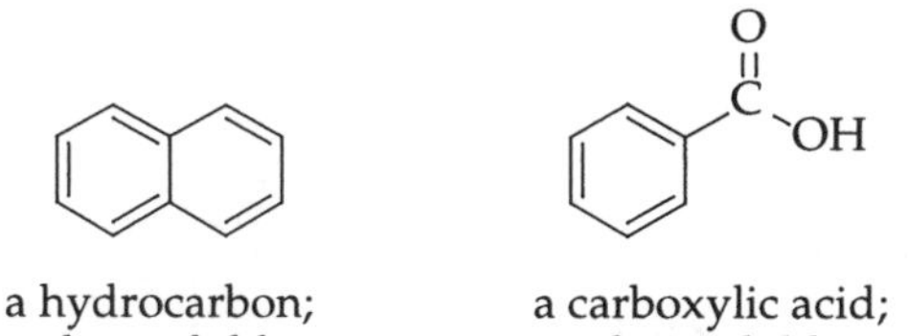

However, if the carboxylic acid is converted to an anion, which will be soluble in water but insoluble in diethyl ether, then we could effect a clean separation of the carboxylic acid (as its anion) from the hydrocarbon. If the above mixture is treated with an aqueous solution of a base, such as sodium hydroxide solution, the carboxylic acid will react to form a water-soluble anion, but the hydrocarbon is neutral and will not react. It will remain water-insoluble and dissolved in the ether layer. By reaction, we have changed the carboxylic acid to an anion and thus have altered its distribution between water and ether. The following illustrates the formation of the water-soluble salt:

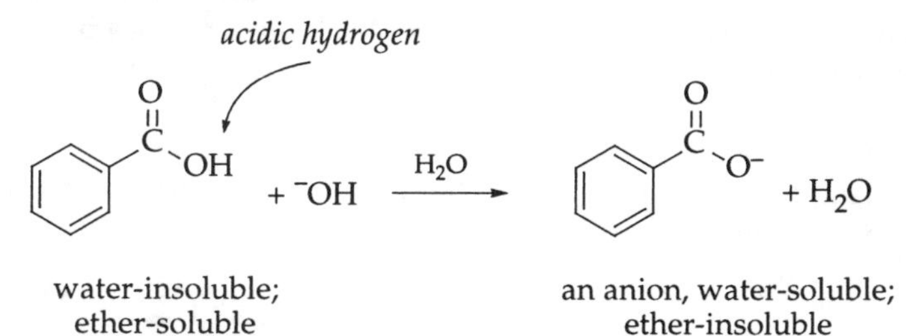

By extraction of an ether solution containing a hydrocarbon and a carboxylic acid with an aqueous solution of NaOH, we can separate the two compounds. The hydrocarbon remains in the ether layer. The carboxylic acid reacts with the NaOH to form an anion, which then leaves the ether layer and dissolves in the water layer. The aqueous layer is separated from the ether layer using a separatory funnel and then acidified to regenerate the original carboxylic acid.

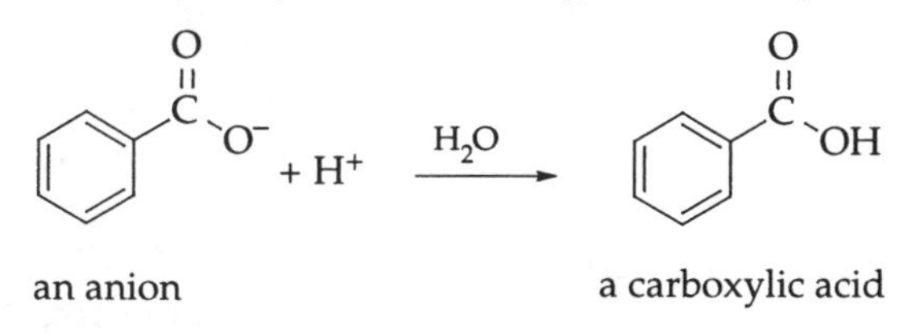

The overall chemical extraction of a hydrocarbon and a carboxylic acid is represented schematically in Figure 3.2.

The preceding discussion illustrates how a carboxylic acid can be separated from a hydrocarbon. The other organic compounds that can undergo an acid–base reaction and thus be separated by chemically active extraction from hydrocarbons are phenols (weak acids) and amines (weak bases). Compounds belonging to these classes can often be separated from other organic compounds because they can be

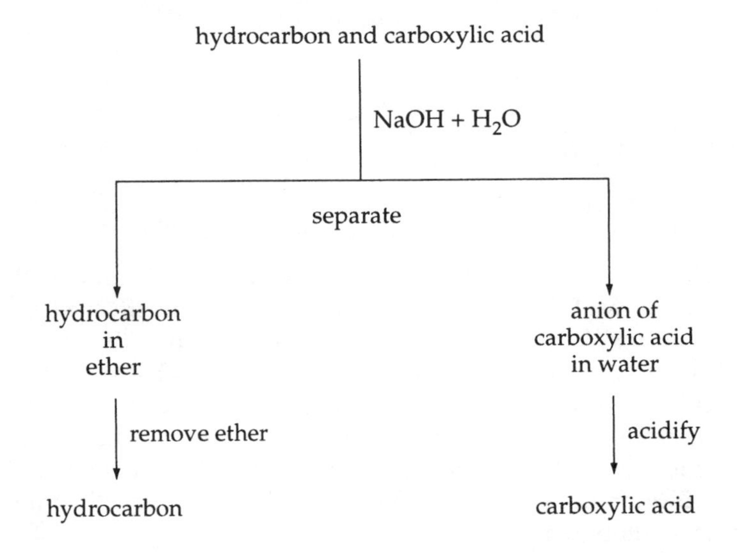

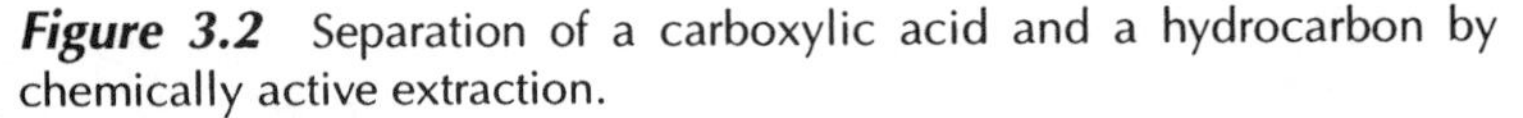

Figure 3.2 Separation of a carboxylic acid and a hydrocarbon by chemically active extraction.

converted to anions by base (in the case of carboxylic acids and phenols) or to cations by acid (in the case of amines).

Weak acids:

RCO_2H $ArCO_2H$ $ArOH$

carboxylic acids pK_a ≅ 5 phenols pK_a ≅ 10

where $-CO_2H$ is $-C(=O)OH$

where Ar = aromatic ring, commonly designated: (benzene ring structures) or (benzene ring structures)

Weak bases:

RNH_2 R_2NH R_3N

amines; pK_a ≅ 4

Carboxylic acids and phenols (but not alcohols, ROH) are strong enough acids to react with a strong aqueous base such as sodium hydroxide to yield water-soluble salts. In this form, these compounds can be extracted from neutral organic compounds.

$$RCO_2H + Na^+ + OH^- \longrightarrow \underbrace{RCO_2^- + Na^+}_{\text{a water-soluble salt}} + H_2O$$

$$ArOH + Na^+ + OH^- \longrightarrow \underbrace{ArO^- + Na^+}_{\text{a water-soluble salt}} + H_2O$$

Carboxylic acids, with pK_a values typically around 5, undergo an acid–base reaction with the weak base sodium bicarbonate ($NaHCO_3$). A typical phenol, with a pK_a value of around 10, is only 1/100,000th as strong an acid as a carboxylic acid. Most phenols are too weakly acidic to undergo reaction with sodium bicarbonate. This difference in reactivity can be used to separate a mixture of a carboxylic acid and a phenol.

$$\underset{\text{water-insoluble}}{RCO_2H} + Na^+ + HCO_3^- \longrightarrow \underbrace{RCO_2^- + Na^+}_{\text{a water-soluble salt}} + H_2O + CO_2\uparrow$$

$$\underset{\text{water-insoluble}}{ArOH} + Na^+ + HCO_3^- \longrightarrow \text{no appreciable reaction}$$

Amines are bases for the same reason that ammonia is a base: An amine contains a nitrogen atom with an unshared pair of electrons and thus can accept a proton. Treatment of an amine with aqueous acid yields a water-soluble salt that can be separated from water-insoluble organic compounds.

$$\underset{\text{water-insoluble}}{RNH_2} + H^+ + Cl^- \longrightarrow \underbrace{RNH_3^+ + Cl^-}_{\text{a water-soluble salt}}$$

Because different compounds form water-soluble salts under different conditions, the acid-base reactions that we have presented can form the basis of a number of types of chemically active extractions. The acid and base solutions commonly employed to effect these reactions are summarized in Table 3.4. Remember that any time a compound exists in an ionic form, it will be soluble in the aqueous phase rather than the organic phase in a separatory funnel. In any of these acid–base reactions, hydrocarbons, halogenated hydrocarbons, alcohols, and other neutral organic compounds are usually unaffected. The separation of an organic acid, an organic base, and a neutral compound by an acid–base extraction is outlined in the flow chart in Figure 3.3.

Table 3.4 Summary of acids and bases used to effect chemically active extractions.

Compound type	*Example*	*Treated with weak base (5% $NaHCO_3$)*	*Treated with strong base (5% NaOH)*	*Treated with acid (5–10% HCl)*
carboxylic acid	C_6H_5–CO_2H	C_6H_5–CO_2^-	C_6H_5–CO_2^-	C_6H_5–CO_2H
phenol	C_6H_5–OH	C_6H_5–OH	C_6H_5–O^-	C_6H_5–OH
amine	C_6H_5–NH_2	C_6H_5–NH_2	C_6H_5–NH_2	C_6H_5–NH_3^+

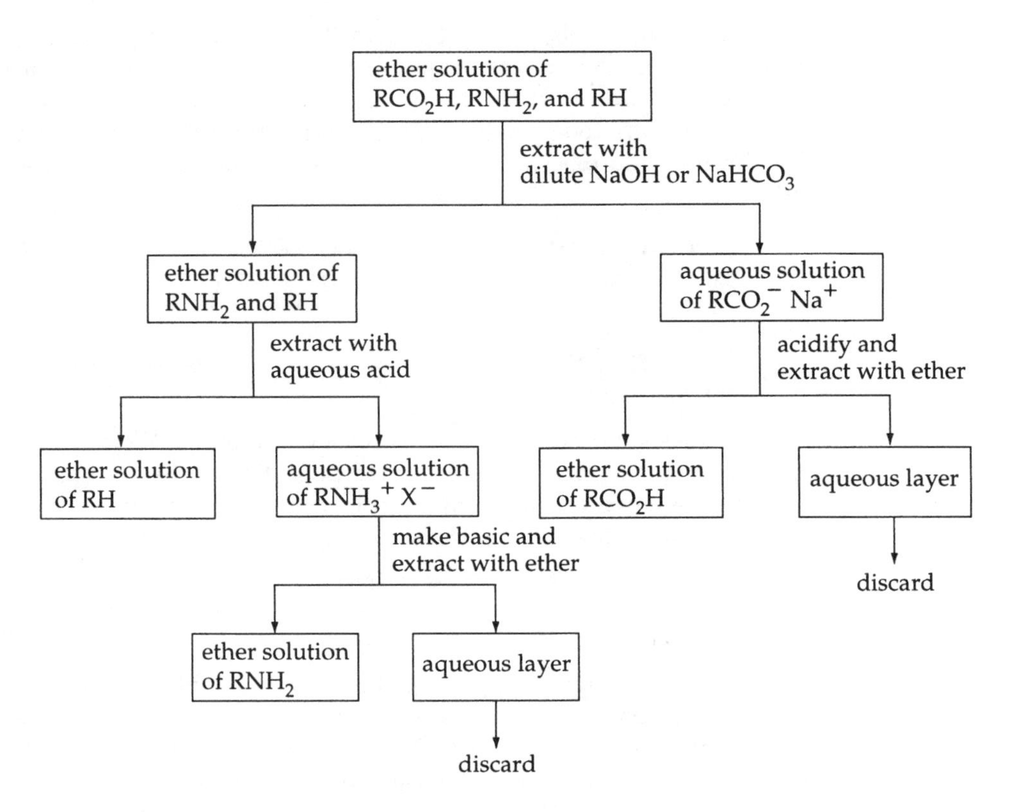

Figure 3.3 Separation of an organic acid (RCO_2H), an organic base (RNH_2), and a neutral organic compound (RH) by an acid–base extraction.

3.5 Steps in Extraction

(1) Preparing the Separatory Funnel

The separatory funnel in your locker might have either a Teflon or a glass stopcock. Place the stopcock in the separatory funnel according to the appropriate instructions in the following two paragraphs.

Teflon stopcock. Make sure that the stopcock is assembled and tightened properly (ask your lab instructor if in doubt). Never grease a Teflon stopcock.

Glass stopcock. Lightly grease the stopcock. Use just enough grease to glaze the clean ground glass when the joint is rotated; too much grease will contaminate the organic solution and may even clog the stopcock. Also, check to be sure the stopcock and funnel are matched. In student laboratories, the separatory funnels and their stopcocks are commonly separated for storage in the storeroom. If the stopcock and funnel are mismatched, the stopcock will leak no matter how well you grease it. You can make a simple check by trying to align the borehole in the stopcock and the stem of the funnel. If they will not align, they are mismatched.

Figure 3.4 shows the separatory funnel as it is positioned in an iron ring just prior to the addition of a liquid.

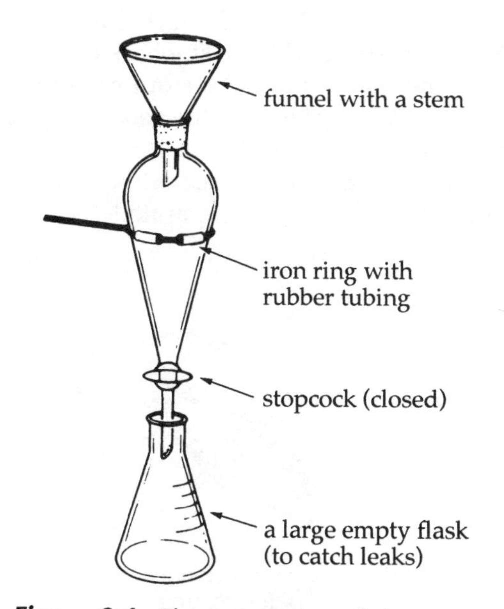

Figure 3.4 The appearance of the separatory funnel before liquids are poured into it.

(2) Adding the Liquids

Be sure the stopcock is closed. Using an ordinary long-stem funnel, pour the solution to be extracted into the separatory funnel, followed by a measured amount of extraction solvent. Do not fill the separatory funnel more than three-quarters full; there will not be enough room for mixing the liquids.

(3) Mixing the Layers

Before inserting the stopper, swirl the separatory funnel gently. The swirling is especially important if an acid and sodium bicarbonate are present, because gaseous carbon dioxide is given off by their reaction. The swirling will drive off some of the carbon dioxide and minimize pressure buildup during the shaking process.

Insert the stopper and, holding the stopper in place with one hand, pick up the separatory funnel and invert it. Immediately open the stopcock with your other hand to vent solvent fumes or carbon dioxide. Swirl the inverted separatory funnel gently with its stopcock open to further drive off solvent vapors or gases. CAUTION: Always aim the stem of the separatory funnel away from people when venting.

Figure 3.5 shows the proper method for holding a separatory funnel. Hold the stopper and stopcock firmly in place throughout the entire shaking process to prevent their falling out.

After venting, close the stopcock, gently shake or swirl the mixture in the inverted funnel, and then revent the fumes. If excessive pressure buildup is not observed, shake the separatory funnel and its contents up and down vigorously in a somewhat circular motion for 2–3 minutes so that the layers are thoroughly mixed.

Vent the stopcock several times during the shaking period. After completing the shaking, vent the stopcock one last time. With the stopcock closed, place the separatory funnel back in the iron ring and remove the stopper. If you are extracting a small amount of material, wash the stopper into the separatory funnel with a few drops of extraction solvent, using a dropper. Place a large Erlenmeyer flask under the stem of the separatory funnel in case the stopcock should develop a leak. Allow the separatory funnel to sit until the layers have separated.

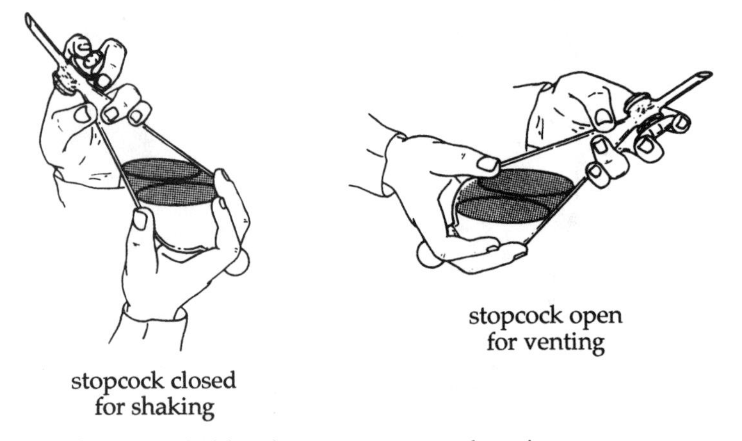

Figure 3.5 How to hold and vent a separatory funnel.

(4) Separating the Layers

Before proceeding, make sure the stopper has been removed. (It is difficult to drain the lower layer from a stoppered funnel, because a vacuum is created in the top portion of the funnel.) Partially open the stopcock to drain the lower layer into the flask. Hold the stopcock in place and brace the separatory funnel with both hands so that the stopcock cannot accidentally slip out of place (see Figure 3.6). Splashing is minimized if the tip of the stem touches the side of the flask so that the liquid can run down its side.

When the lower layer is almost, but not quite, drained into the flask, close the stopcock and gently swirl the separatory funnel. The swirling knocks drops clinging to the sides of the funnel to the bottom, where they can be removed. Carefully and slowly drain the last of the lower layer into the flask. Finally, tap the stem of the funnel to knock any clinging drops into the flask.

At this point in a synthesis experiment, the organic layer containing the desired product would be washed and poured from the top of the funnel into a clean Erlenmeyer flask. In this way, the upper layer is not contaminated with drops remaining in the stem.

(5) Cleaning the Separatory Funnel

Clean the separatory funnel as soon as you are finished so that the organic solvent will not dissolve the stopcock grease and cause the stopper or stopcock to freeze. Then regrease the stopcock (if it is glass) or store it separately. A Teflon stop-

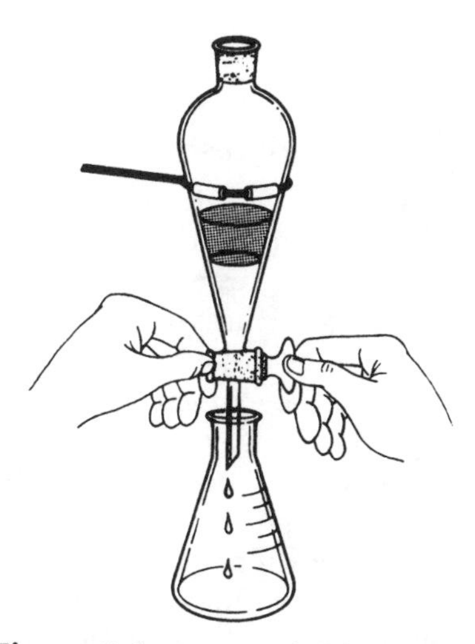

Figure 3.6 How to hold a separatory funnel while draining the lower layer.

cock should be washed, but not greased, and replaced loosely in the separatory funnel.

When the experiment is completed, do not dispose of unwanted extraction layers down the drain. Check with your instructor for the procedure to use to discard chemicals in your laboratory.

Safety Notes

- If you have a separatory funnel with a glass stopcock, make sure that the stopcock and funnel are matched to prevent leaking of potentially harmful solvents.
- Never add a volatile solvent to a warm solution.
- Use caution when mixing any materials that can evolve a gas.
- If you are using a flammable solvent, make sure there are no flames in the vicinity!
- Always aim the stem of the separatory funnel away from your neighbors when venting. Better yet, aim the stem into the hood.
- Organic extraction solvents should be boiled or evaporated only in a fume hood.
- Flammable solvents should never be heated with a burner. A steam bath is the safest source of heat. A spark-free hot plate may also be used.

3.6 Additional Techniques Used in Extractions

A. Microscale and Semi-Microscale Procedures

Separatory funnels, with their sloping sides, well-fitting stopper, and convenient stopcock drainage system are ideal for liquid–liquid extractions. However, the smallest manufactured separatory funnel is too large for the small reaction volumes used in microscale procedures. In these procedures, extractions are carried out in vials or test tubes, as the following paragraphs describe.

Place the solution to be extracted and the extraction solvent in a test tube or a vial. Screw-capped containers work well because they are easy to stopper during mixing. Flat-bottomed conical vials are beneficial both because they do not fall over and because the sloping sides aid in the separation of the layers. Test tubes can also be used; cap them with a snug-fitting cork or rubber stopper. (See Figure 3.7.)

Mix the solution and extraction solvent by flicking the bottom of the container or by stoppering it and shaking it, venting often. Or pipet the contents up and down several times with a Pasteur pipet.

Allow the layers to settle. Then reach into the container with a Pasteur pipet and carefully remove the *lower* layer and place it in a clean container. Depending on the densities of the solvents, the lower layer may be the organic or the aqueous layer. The reason that the lower layer is transferred is that it is very difficult to quantitatively remove the upper layer with a pipet.

As appropriate, the organic layer is subjected to multiple extractions or washing as described in the following sections, all manipulations carried out as illustrated in Figure 3.7 using test tubes or vials.

B. Multiple Extractions

In practice, a single extraction is rarely used; multiple extractions are the rule. A typical double-extraction sequence of an aqueous solution by diethyl ether is diagramed in Figure 3.8.

The original aqueous solution that has already been extracted once and drained into a flask is returned to the dirty, but empty, separatory funnel, along with a fresh portion of extraction solvent. If a second separatory funnel is available, the aqueous layer can be drained directly into it for the second extraction. (If the solvent forms a lower layer, it is drained off first. In this case, the original solution to be reextracted can simply remain in the separatory funnel, and fresh solvent added.) The second (and possibly a third) extraction is carried out just as the first one was. After the multiple extraction, the organic layers are combined in one flask. The aqueous layer (if that is the layer that will be discarded) should be labeled and saved until the entire experiment is finished.

C. Salting Out

If a compound has a low distribution coefficient between an organic solvent and water, one or more simple extractions will not remove much of the compound from the water.

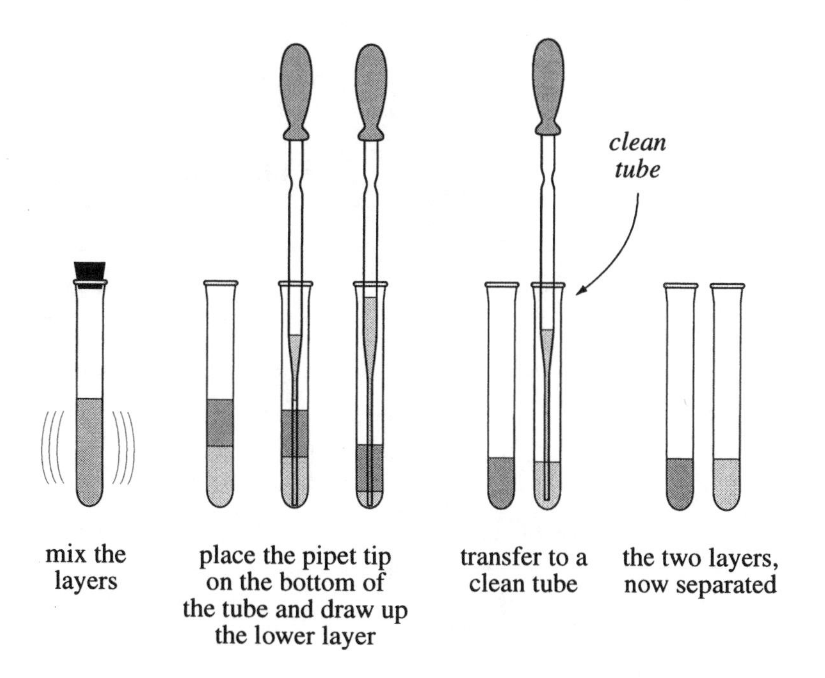

Figure 3.7 Microscale extraction technique using a test tube.

The distribution coefficient of an organic compound between an organic solvent and water can be changed by adding sodium chloride to the water. (Other inorganic salts have the same effect as sodium chloride, but the latter is the least expensive salt available.) Organic compounds are less soluble in salt water than in plain water. Sometimes, the solubility difference is dramatic. Therefore, by simply dissolving sodium chloride in the water layer, we can increase the distribution of an organic compound in the organic solvent. This effect is commonly referred to as "salting out" the organic compound.

D. Emulsions

An **emulsion** is a suspension of one material in another that does not separate quickly by gravity. In liquid–liquid extractions, an emulsion refers to the suspension of one of the solutions in the other (see Figure 3.9). The result is that the two layers do not separate completely, causing a very annoying situation.

If it is known that a particular extraction might lead to an emulsion, a few preventive steps might save time later. The *addition of sodium chloride* to the aqueous layer decreases the solubility of water in the organic solvent, and vice versa, and therefore may prevent an emulsion. *Swirling* the separatory funnel, instead of shaking it vigorously, when mixing the two layers may also prevent an emulsion from forming. However, swirling is a less efficient (and thus slower) way of reaching the equilibrium distribution of solute.

Once an emulsion has formed, it can sometimes be broken up by one or more of the following techniques:

- Allow the separatory funnel to sit in an iron ring for a few minutes; gently

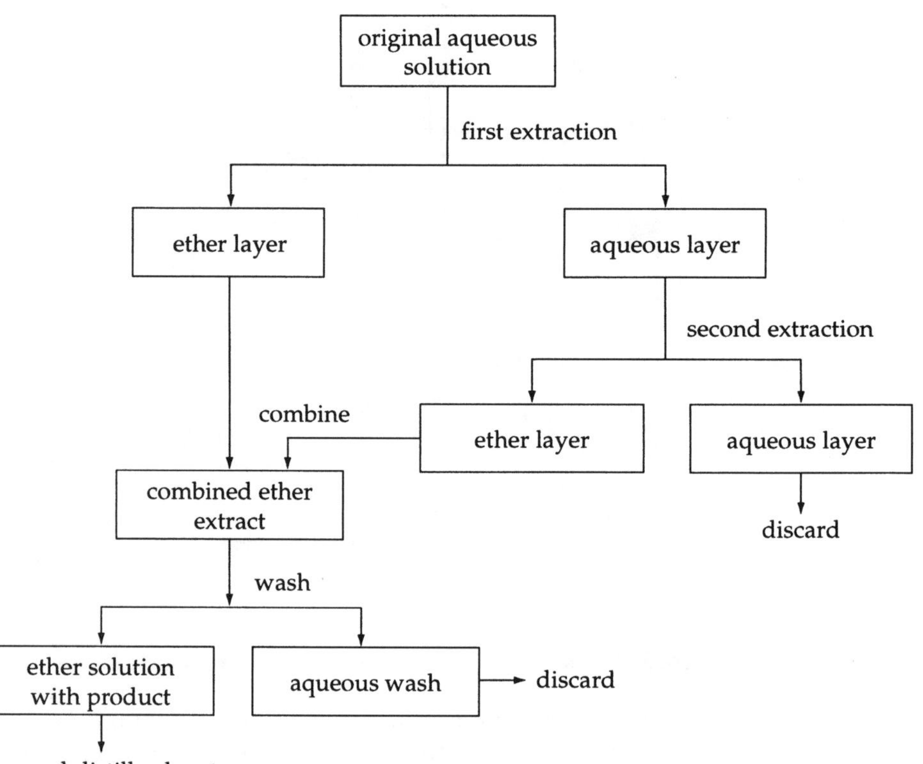

Figure 3.8 Diagram of a two-stage diethyl ether extraction of an aqueous solution.

swirl it periodically.

- Use a dropper to add a few drops of a 64saturated solution of aqueous sodium chloride.
- Squirt a few drops of 95% ethanol or a commercial antifoam agent on the emulsion "bubbles."

If none of the preceding techniques breaks up the emulsion, one of the following techniques might work:

- Filter the mixture with vacuum (which removes solid particles that help stabilize the emulsion), and then proceed with the separation.
- Draw off as much of the lower layer in the separatory funnel as possible, adding fresh solvent to the top layer remaining in the funnel to dilute it, and swirl gently.
- Pour the mixture into a flask, cork the flask, and allow the mixture to sit overnight or until the next laboratory period. (Do not let the mixture stand for any length of time in the separatory funnel; the stopcock will eventually leak under the pressure of liquid.)

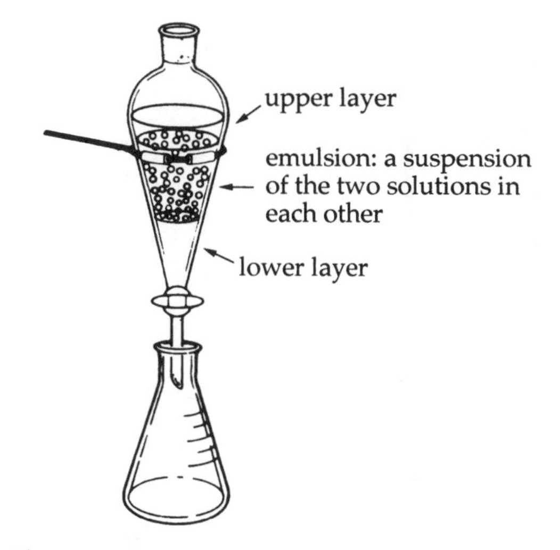

Figure 3.9 An emulsion in the separatory funnel.

E. Washing

If the original solution was aqueous, the organic layers, separately or combined, may be *washed* (extracted with fresh water) to remove any water-soluble impurities. Use a volume of wash water about 10% that of the organic layer. A 5% sodium bicarbonate wash may be used to remove traces of acid from the organic layer. A sodium chloride wash (saturated aqueous solution) may be used to remove water from the organic solution.

For washing, place the organic solution in the separatory funnel along with the wash solution, and then shake and separate in the usual manner. Combine the aqueous washes with the original water solution. Save $NaHCO_3$ or NaCl washes separately. Label and save all extracts until the experiment is complete.

F. Drying and Removal of Solvent

No two liquids are completely immiscible. In any liquid–liquid extraction, the desired layer (usually the organic layer) contains some of the other solvent (usually water). Before the organic solvent is removed, the organic solution should be dried so that water will not contaminate the product. Specific procedures for drying organic solutions are discussed in Technique 4.

After an extraction has been completed and the extract dried, the solvent is removed from the organic compound. There are many ways to do this. Small amounts of solvent can be removed by simple evaporation or by boiling the solvent and using a stream of clean, dry air or vacuum to remove the vapors from the flask. (When heating such a mixture, take care that the residue in the flask does not become overheated.)

Large amounts of solvent should not be boiled away into the atmosphere, even through a fume hood; instead, the solvent should be distilled and collected (see Technique 5).

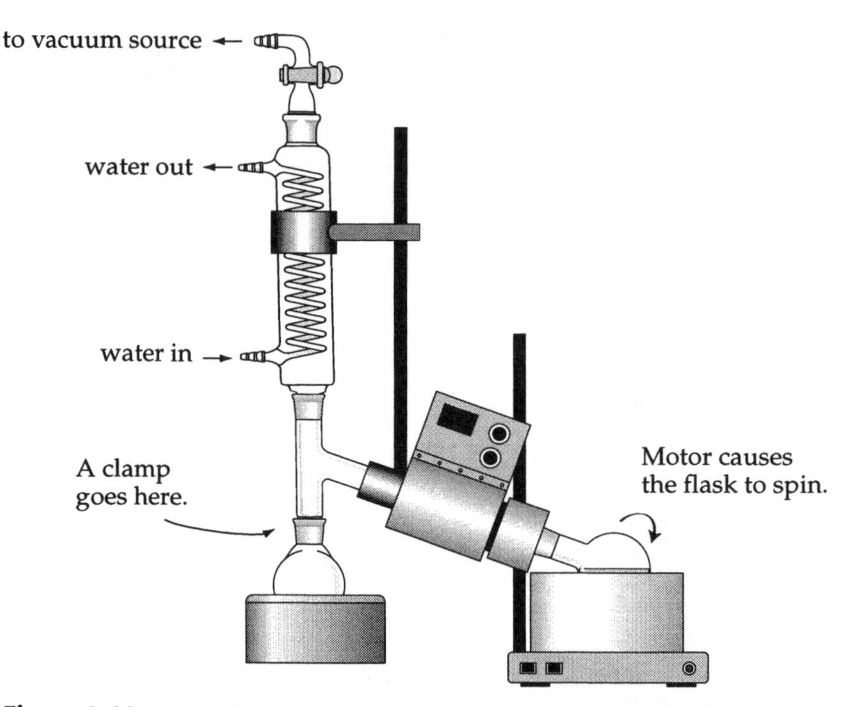

Figure 3.10 A typical roto-evaporator used for the removal of solvent under vacuum.

In some laboratories, you may have access to a rotary evaporator, a convenient vacuum apparatus for removing solvent quickly (see Figure 3.10). If your laboratory is equipped with one of these, your instructor will show you how to use it.

3.7 Other Extraction Techniques

A. Continuous Liquid–Liquid Extraction

The extraction of a compound with a low distribution coefficient between organic solvent and water would ordinarily require large amounts of solvent. The technique of continuous liquid–liquid extraction allows solvent to be recycled through the aqueous solution so that the compound can be completely extracted with a moderate amount of solvent.

Figure 3.11 illustrates a continuous-extraction apparatus that can be used with a lighter-than-water solvent. (Other types of extractors have been designed for heavier-than-water solvents.) The aqueous solution to be extracted is placed in a long tube with a sidearm. Solvent is placed in a distillation flask, as the figure shows. When the solvent is distilled, its condensed vapors drip into a narrow glass tube with a fritted bottom. When the narrow tube is filled with solvent, small bubbles of solvent are forced through the frit to rise through the aqueous solution, extracting organic material as they rise. The organic solution above the water passes through the sidearm back to the distillation flask, where more solvent is distilled. As the desired organic material is extracted from the water, its concentration builds up in the distillation flask.

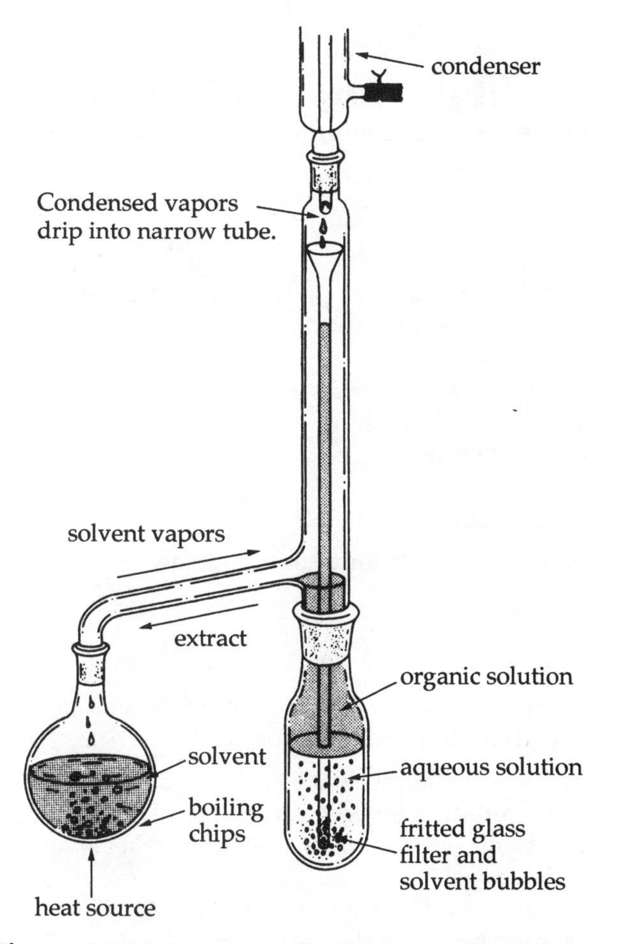

Figure 3.11 An apparatus for continuous liquid–liquid extraction using a solvent that is lighter than water.

A continuous extraction of this sort requires several hours, or even days, but the operator is free to do other activities while the extraction is being carried out. When the extraction is complete, the organic extract is dried and the organic compound is isolated from solution.

B. Continuous Solid–Liquid Extraction

For the continuous extraction of organic compounds from a solid mixture, a Soxhlet extraction apparatus may be used. The solid material is placed in a porous cup, or thimble, made of thick, tough filter paper. The thimble is placed in the inner tube of the extractor. The apparatus allows hot solvent vapors to bypass the thimble through the sidearm, condense, and drip into the thimble. When the solvent fills the thimble, the organic extract siphons through the small siphon tube into the solvent distillation flask. The process is repeated automatically until the extraction is complete. Like continuous liquid–liquid extraction, several hours, or even days, may be required for the extraction.

Additional Resources

A list of suggested readings for the technique of extraction is published on the Brooks/Cole Web site, as well as additional problems. Please visit www.brooks-cole.com.

Problems

3.1 Consider the following solvent pairs. If mixed together, which pairs would form two layers? If they form two layers, which solvent would be on top?
(a) hexanes and water
(b) water and methylene chloride
(c) hexanes and methylene chloride
(d) methanol and hexanes
(e) ethanol and water
(f) acetone and toluene

3.2 The distribution coefficients for a compound are:

benzene-water, $K = 5.5$
petroleum ether-water, $K = 5.0$
methylene chloride-water, $K = 1.0$

Which is the best choice of solvent for extracting the compound from an aqueous reaction mixture? Explain your answer.

3.3 Suppose you added an additional 50 ml of water to a separatory funnel containing a compound distributed between 50 mL of ether and 50 mL of water. (The compound you want is more soluble in ether than it is in water.) How will this addition affect
(a) the distribution coefficient of the compound?
(b) the actual distribution of the compound?

3.4 The pain reliever phenacetin is soluble in cold water to the extent of 1.0 g/1310 mL and soluble in diethyl ether to the extent of 1.0 g/90 mL.
(a) Determine the approximate distribution coefficient for phenacetin in these two solvents.
(b) If 50 mg of phenacetin were dissolved in 100 mL of water, how much ether would be required to extract 90% of the phenacetin in a single extraction?
(c) What percent of the phenacetin would be extracted from the aqueous solution in (b) by two 25-mL portions of ether?

3.5 An aqueous solution containing 5.0 g of solute in 100 mL is extracted with three 25-mL portions of diethyl ether. What is the total amount of solute that will be extracted by the ether in each of the following cases?
(a) Distribution coefficient (ether–water), $K = 0.10$
(b) $K = 1.0$
(c) $K = 10$

3.6 If the compound in problem 3.5 (b) were extracted with three 50-mL portions of diethyl ether, how much would be extracted?

3.7 Complete the following equations. If no appreciable reaction occurs, write "no reaction."

(a) $HO-C_6H_4-CH_3 + NaOH \xrightarrow{H_2O}$

(b) $C_6H_5-CH_2CH_2OH + NaOH \xrightarrow{H_2O}$

(c) $HO-C_6H_4-CH_3 + NaHCO_3 \xrightarrow{H_2O}$

(d) $CH_3CH_2CH_2CO_2H + NaHCO_3 \xrightarrow{H_2O}$

(e) $CH_3NHCH_2CH_2NH_2 + NaHCO_3 \xrightarrow{H_2O}$

(f) $CH_3CH_2CH_2CO_2H + HCl \xrightarrow{H_2O}$

(g) $H_3CH_2C\,C(=O)CH_2CH_3 + HCl \xrightarrow{H_2O}$

(h) $CH_3NHCH_2CH_2NH_2 + HCl \xrightarrow{H_2O}$

3.8 Which of the following pairs of compounds *could* be separated by chemically active extraction? What reagent would you use?

(a) $CH_3CH_2CO_2H$ and $ClCH_2CH_2CO_2H$

(b) $HO-C_6H_4-CH_2CH_2CH_3$ and $HOH_2C-C_6H_4-CH_2CH_3$

(c) NH and O

3.9 Draw a flow diagram for the separation of $CH_3CH_2CH_2Br$, $(CH_3CH_2CH_2)_3N$, and $CH_3CH_2CO_2H$.

3.10 Diagram a two-stage dichloromethane ($d = 1.3$) extraction of an aqueous solution, showing how this procedure differs from the diethyl ether extraction diagrammed in Figure 3.8.

3.11 A reaction workup for an aqueous reaction mixture calls for extraction with diethyl ether and then an extraction with saturated aqueous sodium chloride. What is the purpose of the saturated sodium chloride step?

3.12 You have prepared an organic compound by reacting it with strong aqueous acid. The workup procedure calls for extraction of the compound into methylene chloride, then an extraction of the methylene chloride solution with 5% $NaHCO_3$.

(a) What is the purpose of the extraction with 5% $NaHCO_3$?

(b) What precautions should you take when you add the 5% $NaHCO_3$ to the organic solution?

3.13 What is *wrong* with the following procedure?

"The reaction mixture, consisting of NaBr and an ethanol solution of the product, is diluted with an equal volume of water, then extracted once with an equal volume of diethyl ether. The lower aqueous layer is discarded."

3.14 You carefully purified phenanthrene from a very long, complicated procedure. It took days. It is very important for you to hand the purified phenanthrene in to your instructor in one hour to get a good grade. Oops! You accidentally put the phenanthrene into a vial that you thought was clean but instead had a lot of NaCl clinging to the sides. What can you do to quickly purify the phenanthrene again?

Drying Organic Solutions

Reaction and isolation processes often contaminate organic compounds or solutions with small amounts of water. Various methods for removing water from wet organic solutions are available. The method chosen depends on a number of factors, including the organic compound in solution, the solvent, the degree of wetness, and the degree of dryness desired.

4.1 Extraction with Aqueous Sodium Chloride

A saturated aqueous solution of sodium chloride is an inexpensive drying agent that will remove the bulk of water from a wet organic solution. A final extraction using a separatory funnel with a saturated NaCl solution is especially valuable when the organic extraction solvent is diethyl ether, which can contain 1.2% water at 20°. The salt water works to pull the water from the organic layer to the water layer. This is because the concentrated salt solution wants to become more dilute and because water has a stronger attraction to salts than to organic solvents. (There are other benefits of washing with a saltwater solution; see the section on "salting out," Technique 3, Section 3.6C.) When most of the water is removed by a saturated NaCl wash, the solution can be further dried with a solid inorganic drying agent.

4.2 Solid Inorganic Drying Agents

Solid drying agents are commonly used to remove the last traces of water from organic solutions. These drying agents are generally anhydrous inorganic salts (insoluble in organic liquids) that absorb water and are converted to hydrated salts. Molecular sieves are aluminosilicates that contain cavities that can trap molecules of certain sizes and shapes. Some types of molecular sieves are excellent drying agents.

Table 4.1 lists a few common drying agents and comments on their use. Most drying agents are reasonably swift (15 minutes) in removing the bulk of the water from an organic solution; however, most are quite slow in trapping the last vestiges of water. For this reason, an overnight drying is always preferable to a 15-minute drying. If a trace of water is tolerable, however, the 15-minute drying may allow you to continue the experiment and save laboratory time.

The amount of water a drying agent can remove from a wet solvent depends on the solvent and the structure and stability of the hydrate. For example, calcium sulfate, which forms the hydrate $CaSO_4 \bullet \frac{1}{2} H_2O$, cannot remove as much water as

Table 4.1 Some drying agents for organic solutions.

Name	*Formula*	*Speed*	*Practical capacity**	*Comments*
molecular sieve 4A (powder)†	—	very fast	very high	neutral and easy to use, but expensive
calcium chloride	$CaCl_2$	fast	high	forms complexes with O and N compounds, such as alcohols, amines, ketones, and carboxylic acids; may contain CaO as an impurity
magnesium sulfate	$MgSO_4$	moderate	high	best filtered, not decanted, because of fine particle size
potassium carbonate	K_2CO_3	moderate	moderate	basic—reacts with acidic compounds such as phenols and carboxylic acids
calcium sulfate	$CaSO_4$	fast	low	neutral; available with indicator
sodium sulfate	Na_2SO_4	fast	low	neutral; easy to use
saturated sodium chloride solution	$NaCl + H_2O$	very fast	low	used to remove the bulk of water from an organic solution so that less solid drying agent is needed

* Based on the quantity of water removed from wet ether per gram of drying agent.
† Molecular sieves are also available as beads, which are not as fast or efficient as the powdered form.

can calcium chloride, which can form hydrates containing 1–6 moles of water per mole of $CaCl_2$.

Your choice of drying agent will depend on the compound you are drying, the drying time available, the degree of dryness necessary, and the expense. For everyday classroom use, anhydrous calcium chloride, magnesium sulfate, and sodium sulfate are the most useful. Table 4.1 lists the limitations and benefits of these drying agents. Note that calcium chloride should not be used for drying compounds containing unshared valence electrons, such as alcohols and amines.

In special cases, drying agents that actually react with water are used. For example, diethyl ether is often dried over sodium metal. Very wet ether cannot be dried this way because of the violent reaction of sodium with water ($2\ Na + 2\ H_2O \rightarrow 2\ NaOH + H_2$). Ether solutions of organic compounds are also not dried over sodium. Calcium hydride is another example of a drying agent that reacts quite vigorously with water ($CaH_2 + H_2O \rightarrow CaO + 2\ H_2$).

4.3 Procedure for Using Solid Inorganic Drying Agents

A moderate amount of drying agent (about 1–5 g, or just enough to cover the bottom of the flask) is sufficient for drying most solutions. Add the drying agent, cork the flask, and swirl the contents of the flask to speed the drying. If the drying agent becomes wet-looking or clumped, filter or decant the solution into a clean, dry flask and add a fresh portion of drying agent. Finally, remove the hydrated salts by gravity filtration (see p. 31). The hydrated salts can also be removed by carefully decanting the organic solution into a clean flask. The residual organic solution can be transferred by using a disposable pipet. (See Figure 1.4, p. 34.)

Often the procedure detailed in the above paragraph is abbreviated "dry over" in an experimental procedure.

4.4 Azeotropic Drying

Some solvents form **low-boiling azeotropes** with water—that is, when they are distilled, they produce a mixed distillate of constant composition with a lower boiling point than that of either water or the pure solvent (see Technique 5, Section 5.1D). For example, benzene (bp 80.1°) and water (bp 100.0°) form an azeotrope composed of 91% benzene and 9% water that boils at 69.4°. Table 4.2 lists some solvents that form low-boiling azeotropes with water.

Table 4.2 Composition and boiling points of some low-boiling binary azeotropes containing water.

Composition	*bp of azeotrope (°C)*
91% benzene (bp 80.1°) 9% water (bp 100.0°)	69.4
96% carbon tetrachloride (bp 76.8°) 4% water	66.8
97% chloroform (bp 61.7°) 3% water	56.3
99% dichloromethane (bp 41°) 1% water	38.8

Solutions in these solvents may be dried by simply distilling the azeotropic mixture until the distillate is clear and no longer a two-phase mixture of water plus solvent. Once the azeotrope is distilled, the residual solution contains no more water.

A Dean-Stark trap is a device used to remove water from a reaction mixture or solution by azeotropic distillation. To use a Dean-Stark trap, assemble the device as shown in Figure 4.1. The organic solution to be dried (or reaction mixture from which water is to be removed) is placed in the round-bottom flask. The organic solvent must be one that forms azeotropes with water. In addition, the solvent used for the azeotropic drying must be less dense than water. Benzene is a suitable solvent for use in a Dean-Stark apparatus.

The solution to be dried is boiled, causing the azeotrope to distill and be condensed in the condenser. Upon condensation, the azeotrope separates into two layers—the organic solvent layer and the water layer. Both layers drain into the Dean-Stark trap. The heavier water sinks to the bottom of the trap. The lighter solvent floats on top of the water and, when it fills the trap, drains back into the flask.

The Barrett receiver (see Figure 4.1) is similar to a Dean-Stark trap. The difference is that the Barrett receiver contains a stopcock at the bottom of the water trap. This allows the periodic removal of water as it accumulates. The major disadvantage of the Barrett receiver is that the stopcock tends to leak when used for an extended period of time.

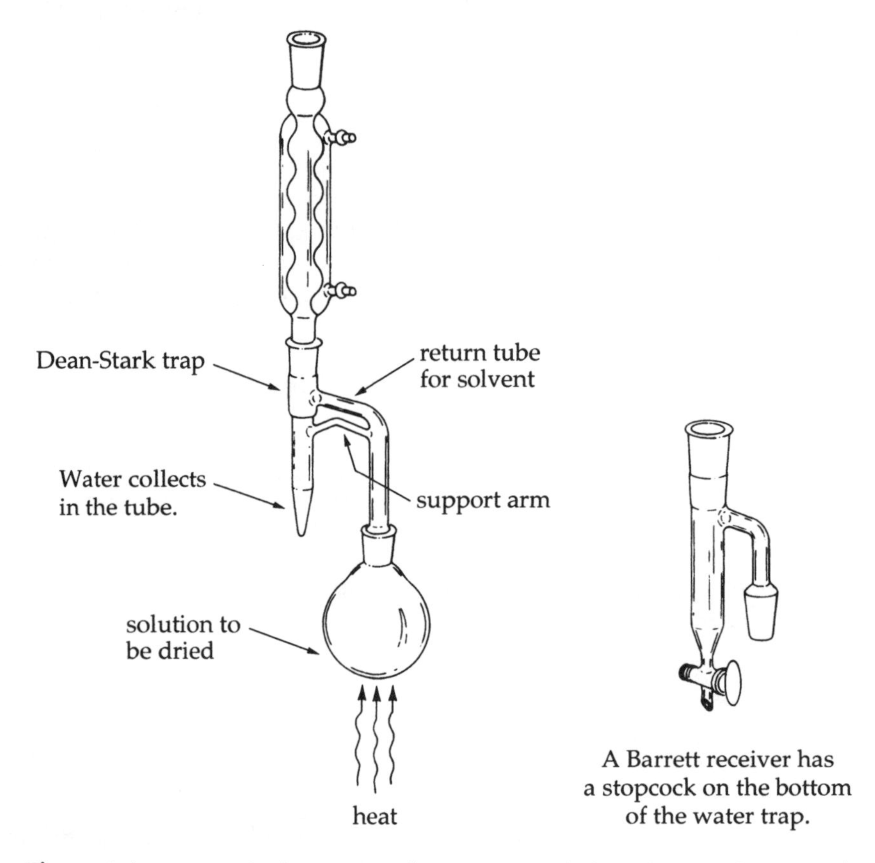

Figure 4.1 A Dean-Stark apparatus for azeotropic drying. (Source: Courtesy of VWR Equipment Catalogue.)

Additional Resources

A list of suggested readings for the technique of drying organic solutions is published on the Brooks/Cole Web site, as well as additional problems. Please visit www.brookscole.com.

Problems

4.1 Which drying agents in Table 4.1 could be used to dry an ether solution of each of the following compounds?

(a) $CH_3CH_2CH_2CH_2OH$

(b) $CH_3CH_2CH_2CH_2Br$

(c) $CH_3CH_2CO_2H$

(d) $CH_3CH_2CH_2CH_2NH_2$

4.2 Why must a solid drying agent be removed from a solution before the solution is distilled?

4.3 Because anhydrous sodium sulfate is a relatively inefficient drying agent, a student uses twice as much of this salt as recommended. What are the advantages and disadvantages of adding extra drying agent?

4.4 A student adds too much drying agent to an organic solution and the organic solution is completely covered by the drying agent. Explain how he or she can salvage the organic solution.

4.5 A solution of an organic compound in diethyl ether is dried over sodium sulfate (Na_2SO_4). Why should the hydrated salts be removed by gravity filtration, and not the faster method of vacuum filtration?

4.6 A solution of methylene chloride is dried by extracting with a solution of saturated sodium chloride solution, using a separatory funnel. When the two layers are allowed to separate, which layer will float to the top?

Technique 5

Simple Distillation

Distillation is a general technique used for removing a solvent, purifying a liquid, or separating the components of a liquid mixture. In distillation, a liquid is vaporized by boiling, then condensed back to a liquid, called the **distillate** or **condensate**, and collected in a separate flask (the receiving flask). In an ideal situation, a low-boiling component can be collected in one flask, and then a higher-boiling component can be collected in another flask, while the highest-boiling components remain in the original distillation flask as the **residue**.

5.1 Characteristics of Distillation

A. The Boiling Point

The **boiling point** of a liquid is defined as the temperature at which its vapor pressure equals the external pressure acting on the surface of the liquid. The external pressure is usually the atmospheric pressure. For instance, consider a liquid heated in an open flask. The vapor pressure of the liquid will increase as the temperature of the liquid increases, and when the vapor pressure equals the atmospheric pressure, the liquid will boil. Different compounds boil at different temperatures because each has a different, characteristic vapor pressure; compounds with higher vapor pressures will boil at lower temperatures.

The distillation method of boiling-point determination measures the temperature of the vapors above the liquid. In theory, the temperature of the vapor and the boiling liquid are the same since the two are in equilibrium. However, the vapor temperature rather than the pot temperature is measured because if you put a thermometer actually in the boiling liquid mixture, the temperature reading would likely be higher than that of the vapors. This is because the liquid can be superheated or contaminated with other substances, and therefore its temperature is not an accurate measurement of the boiling temperature.

Figure 5.1 is the graph of the vapor pressure of acetone versus its temperature. Because a liquid boils when its vapor pressure equals the pressure above it, the term *vapor pressure* in the figure actually means "applied pressure." Acetone boils at 56.2°C when the applied pressure is 760 mm Hg (1.0 atm or 760 torr). If the applied pressure is less, for instance, 500 mm, the boiling point of acetone is only 45°C. For this reason, the boiling point is usually reported at a particular pressure: "acetone, bp 45°C (500 mm Hg)."

Figure 5.7, p. 85, shows the apparatus for a **simple distillation**. When the liquid in the distillation flask boils, vapor rises to the top of the flask, through the dis-

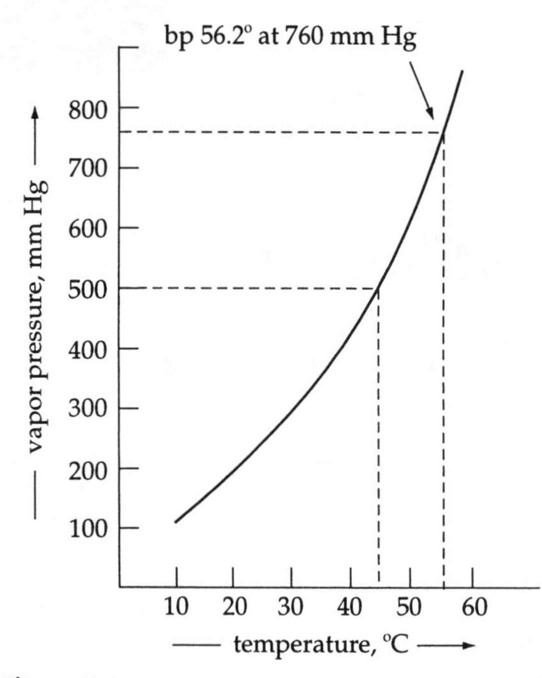

Figure 5.1 Vapor pressure–temperature diagram for acetone, $(CH_3)_2C{=}O$.

tillation head, past the thermometer, and out the sidearm into the condenser. Some of the vapor condenses on the walls of the flask and head and on the thermometer. As long as vapor is flowing out the sidearm and the thermometer bulb is immersed in the vapor, the condensed liquid on the tip of the thermometer is in equilibrium with the vapor, and an accurate boiling point can be determined. Like the melting point, a boiling point is usually reported as a range. A pure compound exhibits a range of 1°–2° or less (but not all liquids with constant boiling points are single pure compounds; see Section 5.1D).

B. Distillation of a Single Volatile Liquid

When a solvent containing a nonvolatile component is removed from a solution by heating, or when a reasonably pure liquid is distilled, the observed temperature rises rapidly to the boiling point. Once the distillation apparatus has reached thermal equilibrium, the boiling point remains relatively constant, not changing more than 1°–2° during the course of the distillation. A significant drop in temperature is the signal that the distillation of the solvent or the pure liquid is complete (see Figure 5.2).

C. Distillation of Mixtures

Because most distillations are performed with mixtures, let us consider the theory of distillation of an ideal (noninteracting) mixture of two miscible liquids A and B, where A is the lower-boiling component. If the difference in boiling points between A and B is large (100° or more), then the distillation temperature will rise to the boil-

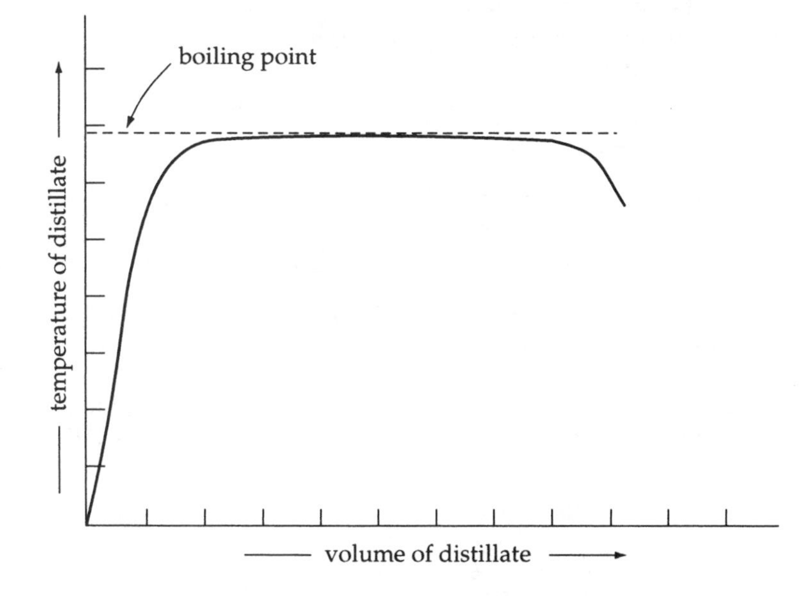

Figure 5.2 Volume versus temperature of the distillate in the distillation of a pure liquid.

ing point of compound A, the lower-boiling component, and remain constant while A is distilling. When almost all of compound A has distilled, the temperature will begin to rise rapidly toward the boiling point of compound B, the higher-boiling component. Once it reaches the boiling point of B, it will remain constant while B is distilling. Unfortunately, this type of distillation is rarely encountered in laboratory practice, because separations usually involve mixtures of compounds with boiling points closer together than 100°.

A more common experience is that the distillation temperature rises more or less steadily during the distillation because *mixtures* of A and B distill. The first, lower-boiling portions of the distillate contain more of A than of B, but this distillate will not be pure A. Similarly, the last, higher-boiling portions of the distillate will be predominantly B but will also contain some A. Although enrichment of A and B in the first and last portions of the distillate can be accomplished, neither pure A nor pure B will be obtained. The reason is that B has a significant vapor pressure, even at temperatures below its boiling point. The vapor pressure of B in a boiling mixture of A and B is a function of a number of factors. To discuss these, we must consider Dalton's law and Raoult's law.

Dalton's law and Raoult's law. A liquid boils when its vapor pressure equals the atmospheric pressure. **Dalton's law of partial pressures** states that the total pressure of a gas is the sum of the partial pressures of its individual components. Thus, the total vapor pressure of a liquid (P_{total}) is the sum of the partial vapor pressures of its components. In our example, the mixture of A and B boils when the sum of the two partial pressures (P_A and P_B) equals the atmospheric pressure.

$$P_{total} = P_A + P_B$$

The **mole fraction X**, a measure of concentration, is the number of moles of one particular component in a mixture divided by the total number of moles (all components) present.

$$X_A = \frac{\text{moles of A}}{\text{moles of A + moles of B}}$$

$$X_B = \frac{\text{moles of B}}{\text{moles of A + moles of B}}$$

where X_A and X_B are the mole fractions of A and B in the mixture

Raoult's law states that, at a given temperature and pressure, the partial vapor pressure of a compound in an ideal solution is equal to the vapor pressure of that pure compound multiplied by its mole fraction in the liquid. By Raoult's law, if $P_A°$ represents the vapor pressure of *pure* A at a specific temperature, then the partial vapor pressure of A (P_A) in a solution is equal to $X_AP_A°$. Similarly, the partial vapor pressure of B (P_B) at that temperature is $X_BP_B°$.

For the liquid mixture:

$$P_A = X_AP_A° \quad \text{and} \quad P_B = X_BP_B°$$

And because $P_{total} = P_A + P_B$, then

$$P_{total} = X_AP_A° + X_BP_B°$$

Raoult's and Dalton's laws state mathematically the rather intuitive concept that the vapor composition above a liquid mixture is dependent both on the vapor pressures of the pure components and on their mole fraction in the mixture. The higher the vapor pressure of a liquid, the more vapors of this component will be above the liquid. And, the greater the concentration of a liquid component, the greater the concentration of its vapors above the liquid.

Composition of the liquid. Distillation is a dynamic process. Vapor is removed as condensed distillate, while more vapor is generated by the boiling liquid. During the course of distillation, the liquid mixture of A and B becomes progressively poorer in the lower-boiling component A and richer in the higher-boiling component B.

To illustrate how the composition changes during a distillation, let us consider what happens to a mixture of A (bp 80.1°) and B (bp 110.6°) in a molar mixture of 1:1 ($X_A = 0.50$ and $X_B = 0.50$). Compound A boils at a lower temperature than does compound B; therefore, $P_A°$ is larger than $P_B°$. For this reason, compound A initially contributes a larger partial vapor pressure ($P_A = X_AP_A°$). At the start of the distillation, the vapor from the boiling mixture contains more A than B. As the distillation proceeds, the boiling liquid contains progressively less A: X_A decreases and X_B increases.

Because $P_B°$ is lower than $P_A°$, the temperature of the liquid must be increased to maintain a boil as the liquid becomes richer in B. Figure 5.3 is a plot of the relative concentrations of A and B versus the temperature of the boiling liquid. In a distillation, we would start at one point on the curve and move to the right.

At the start of the distillation of a 1:1 mixture of A and B, the vapor contains a greater mole fraction of A because P_A is greater than P_B. As the distillation proceeds,

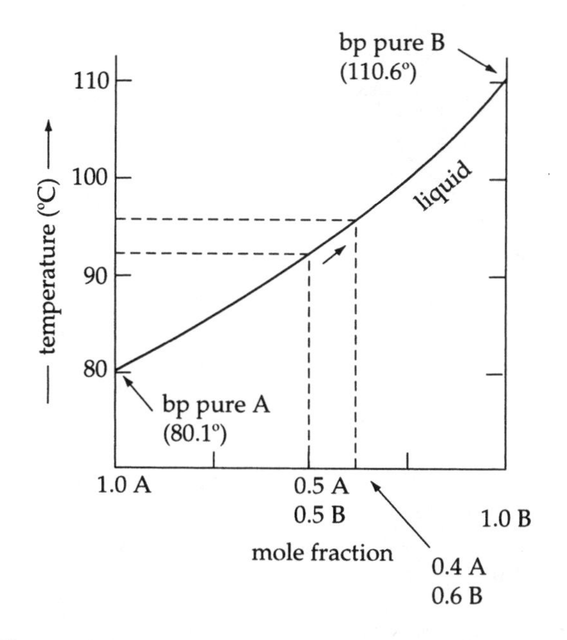

Figure 5.3 The boiling point of an ideal binary liquid solution versus mole fractions of components in the liquid. (The arrow on the curve shows the direction of change in temperature and composition as the distillation proceeds.)

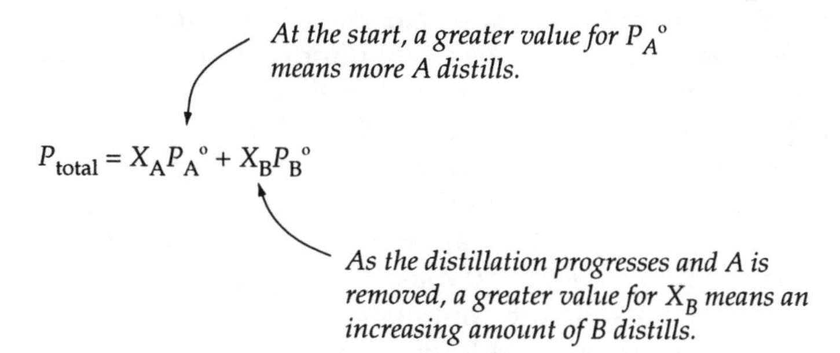

At the start, a greater value for $P_A°$ means more A distills.
$P_{total} = X_A P_A° + X_B P_B°$
As the distillation progresses and A is removed, a greater value for X_B means an increasing amount of B distills.

P_A decreases because of the decreasing amount of A in the liquid, and thus X'_A decreases.

Composition of the vapor. The mole fraction of a compound such as A in the vapor above a boiling mixture (not in the boiling liquid itself) is equal to the ratio of its partial vapor pressure (P_A) to the total pressure.

For the vapor:

$$X'_A = \frac{P_A}{P_{total}} \quad \text{and} \quad X'_B = \frac{P_B}{P_{total}}$$

where X'_A and X'_B are the mole fractions of A and B in the vapor.

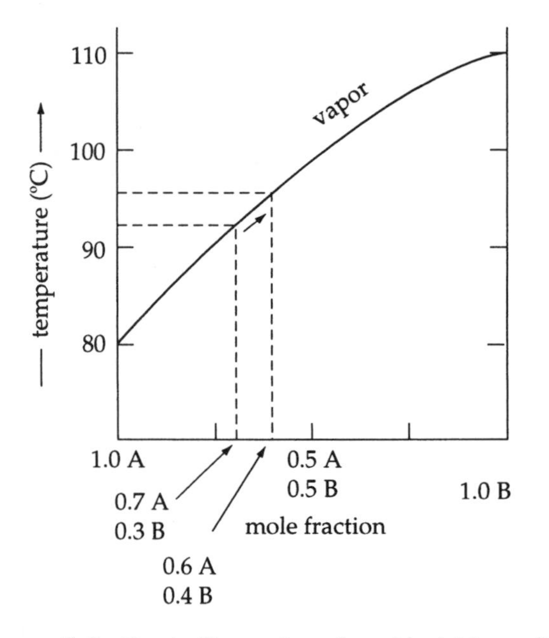

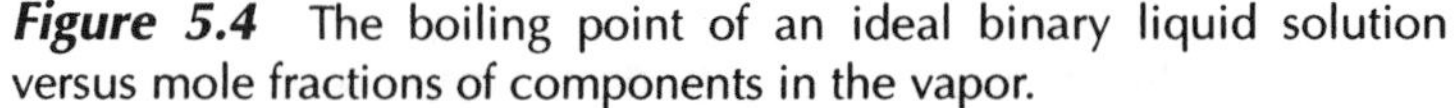

Figure 5.4 The boiling point of an ideal binary liquid solution versus mole fractions of components in the vapor.

Figure 5.4 is a plot of the vapor composition versus the temperature of the vapor. The temperature of the vapor gradually rises as the amount of B in the vapor increases.

Relating liquid and vapor compositions. Generally, the curves for liquid and vapor compositions versus temperature are combined into a single diagram called a *boiling-point-composition diagram,* or *phase diagram,* as shown in Figure 5.5 (a combination of the two curves previously presented in Figures 5.3 and 5.4). In Figure 5.5, we show the same diagram twice to show what happens in the distillation of a 1:1 mixture of A and B. From the first diagram, we see that a 1:1 mixture of A and B boils at 92° and that the vapor at 92° contains approximately 70 mole % A and 30 mole % B (mole fractions 0.7 and 0.3, respectively).

As the distillation continues, more A is distilled than B; therefore, the liquid contains a progressively larger percentage of B. When the boiling liquid contains 40% A and 60% B (mole fractions 0.40 and 0.60, respectively), its temperature is 95°. At 95°, however, the vapor contains 60% A–40% B (mole fractions 0.6 and 0.4). These compositions are marked in the second diagram in Figure 5.5. As the distillation continues, we move farther to the right on each curve. The net results of the changing compositions are (1) a steady increase in the boiling point, and (2) a distillate containing progressively less A and more B.

In our example, a distillate of pure A could never be obtained. However, the early portion of the distillate (containing 80% or 90% A, for example) could be redistilled, yielding somewhat purer A at the start of the distillation. Then, *this* distillate could be redistilled, yielding even purer A. Because these distillations would be tedious and time-consuming, *fractional distillation* was developed. We will discuss fractional distillation as Technique 6.

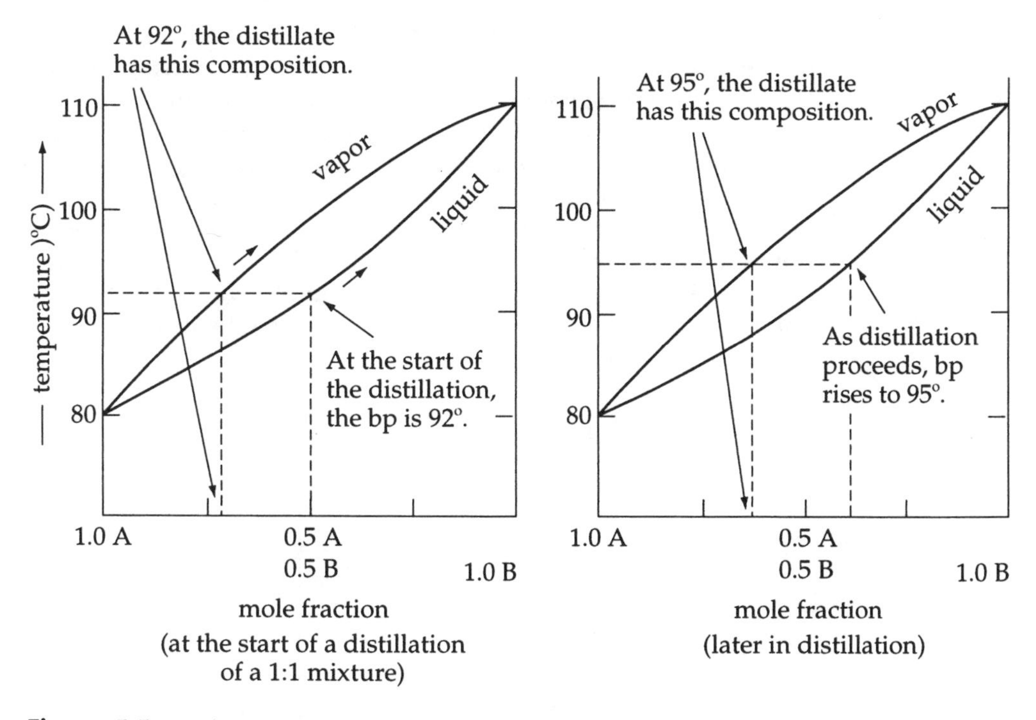

Figure 5.5 Boiling-point-composition diagram, or phase diagram, for an ideal binary solution. (A distillation of a mixture of A and B would begin at the liquid of that composition and progress from there to the right along the lower curve. The composition of the vapor at each temperature is read from the upper curve.)

D. Azeotropes

Because of intermolecular interactions, such as hydrogen bonding, many binary mixtures do not follow Raoult's law and do not have idealized phase diagrams, such as the ones shown in Figure 5.5. One example is a pair of liquids that forms an **azeotrope**, which is a mixture that distills at a constant boiling point and with a constant composition. The phase diagram for one type of azeotrope is depicted in Figure 5.6. The low point on the graph represents the azeotropic mixture, which boils at a constant boiling point, just as a pure compound does. Note that the boiling point of the azeotrope shown in Figure 5.6 is lower than that of either pure component. Because the boiling point is lower, the azeotrope will distill before a component present in excess. The excess component will not distill as a pure compound until the azeotrope has completely distilled.

Although most azeotropes have boiling points lower than those of their components, some azeotropes have boiling points intermediate between those of their components, while other azeotropes have higher boiling points than those of the components. For example, the azeotrope of α-bromotoluene (bp 183.7°) and *n*-octanol (bp 195°) boils at an intermediate temperature of 184.1°, and the azeotrope of acetone (bp 56°) and chloroform (bp 61°) boils at a higher temperature of 64.7°. Azeotropes of two, three, or even more components are also well known.

A common azeotropic mixture encountered in the laboratory is the binary azeotrope of ethanol (CH_3CH_2OH) and water (for others, see Table 4.2 on p. 73). For use in beverages, ethanol is made by the fermentation of aqueous sugars and starches. The fermentation of the carbohydrate mixture stops when the alcohol content of the mixture reaches 12%–14% by volume, because ethanol at this concentration inhibits the action of the enzymes. To yield a higher concentration of alcohol, the fermentation mixture is distilled. Ethanol (bp 78.3°) and water (bp 100°) form a low-boiling azeotrope (bp 78.15°); therefore, this is the first major fraction to distill. This azeotrope contains 95% ethanol and 5% water. After all the ethanol has distilled as part of the azeotrope, the remaining water distills at its normal boiling point. Pure ethanol cannot be distilled from an ethanol–water solution that contains 5% or more water.

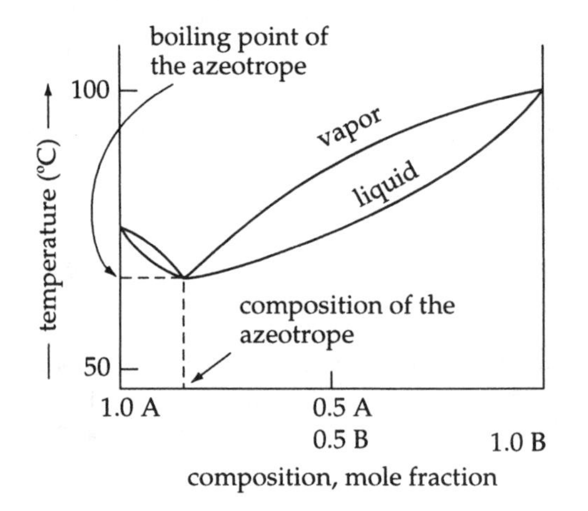

Figure 5.6 Phase diagram for a typical binary mixture that forms a low-boiling azeotrope.

5.2 Steps in a Simple Distillation

The apparatus for a simple distillation is shown in Figure 5.7. Study this figure carefully, noting the placement and clamping of the distillation flask, the distillation head, and the condenser. Note that water flows into the bottom of the condenser's cooling jacket and out the top. If the water inlet were at the top, the condenser would not fill. Also, note the placement of the thermometer bulb, *just below the level of the sidearm* of the distillation head. If the bulb were placed higher than this position, it would not be in the vapor path and consequently would show an erroneously low reading for the boiling point.

(1) The Distillation Flask

Use only a round-bottom flask, never an Erlenmeyer flask, for distillation. The flask should be large enough that the material to be distilled fills 1/2 to 1/3 of its volume. If the flask is overly large, a substantial amount of distillate will be lost as vapor filling the flask at the end of the distillation. If the flask is too small, boiling

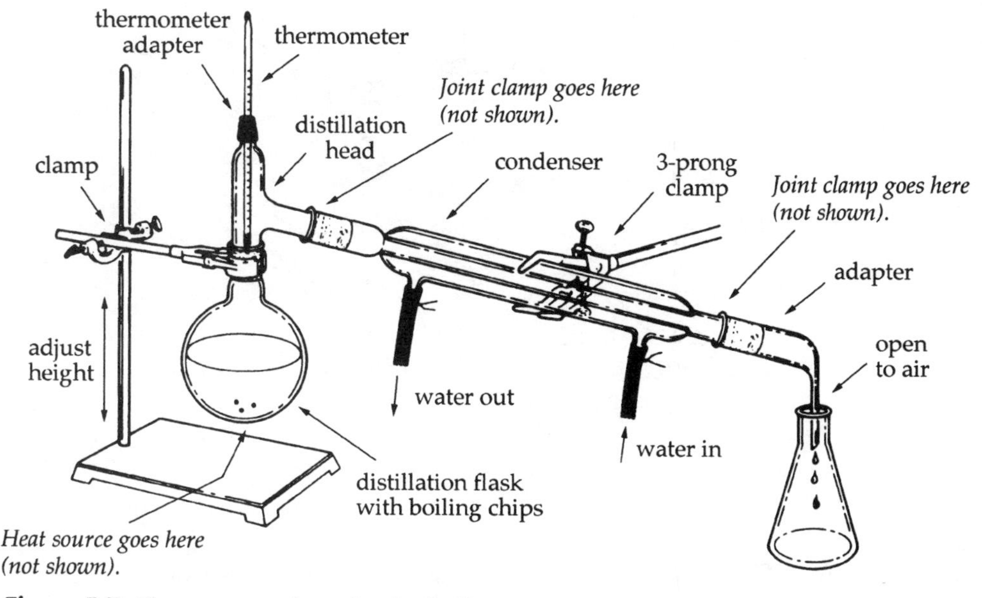

Figure 5.7 The apparatus for a simple distillation.

material may foam, splash, or boil up into the distillation head, thus ruining the separation.

Grease the ground-glass joint of the flask lightly; then securely clamp the flask to a ring stand or rack. Before adding liquid, support the bottom of the flask with a heating mantle on an iron ring (see Figure 5.8). A soft heating mantle should fit snugly around the flask. A hard mantle must fit, or be slightly larger than, the flask. Pour the liquid into the distillation flask, using a funnel with a stem to prevent the liquid from contaminating the ground-glass joint. Finally, add two or three boiling chips.* (CAUTION: Never add boiling chips to a hot liquid!)

(2) The Distillation Head

Grease the ground-glass joints of the distillation head lightly and place the head on the flask, rotating the joints to disperse the grease. It is usually not necessary to clamp the head. (Do not attach the thermometer at this time.)

(3) The Condenser

Grease the ground-glass joints of the condenser lightly and attach rubber tubing firmly to the jacket inlet and outlet (which should not be greased). A strong clamp (oversized, if available) is needed to hold the condenser in place. The weight and angle of the condenser will tend to pull it away from the distillation head; therefore, check the tightness of this joint frequently before and during a distillation. Plastic joint clamps or spring clamps can be used for holding these joints together but not for supporting weight.

---

* The use of boiling chips is discussed in Technique 14, p. 155.

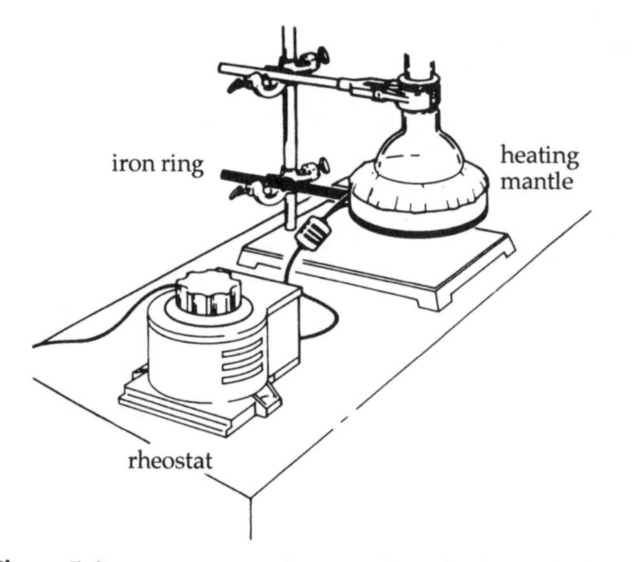

Figure 5.8 Heat source and support for a distillation flask.

Attach the rubber tubing from the lower end of the condenser to an adapter on the water faucet. Place the tubing from the upper end of the condenser in the sink or drainage trough. Turn on the water cautiously. Water should fill the condenser and gently flow, not drip, from the outlet tubing; however, a forceful flow of water will cause the tubing to pop off the condenser. To help prevent this, twist short pieces of wire around the tubing on the condenser inlet and outlet. Because water pressure varies and faucets tend to tighten gradually, check the flow of water frequently during the distillation.

(4) The Adapter

The adapter directs the flow of distillate plus uncondensed vapors into the receiving flask. Figure 5.7 shows the usual type of adapter, but a vacuum adapter may also be used. If desired, a piece of rubber tubing attached to the vacuum adapter can be used to carry fumes to the floor. (Rubber tubing is no substitute for a fume hood, however.) Whichever type of adapter is used, grease its joint lightly before attaching it to the condenser. Plastic joint clamps can be used to secure the adapter to the condenser.

(5) The Receiving Flask

Almost any container can be used as a receiver, as long as it is large enough to receive the expected quantity of distillate. An Erlenmeyer flask is recommended. A beaker is not recommended because its wide top allows vapors and splashes to escape and allows dirt to get into the distillate. Either set the receiving flask on the benchtop or clamp it in place. (It is not good practice to prop up a receiving flask on a stack of books.) If you are collecting several fractions, prepare a series of clean, dry, tared (weighed empty) flasks. If the volume, rather than the weight, of distillate is to be determined, you may use a clean, dry graduated cylinder as the receiver. A round-bottom flask with a ground-glass joint is also a good receiver. This type of flask will fit onto a vacuum adapter, but it must be clamped in place.

If a volatile compound is being distilled, the receiver can be chilled in an ice bath to minimize loss by vaporization (Figure 5.9). A slow distillation with an efficient condenser (keeping a moderately fast water flow at all times) also helps minimize loss of the compound.

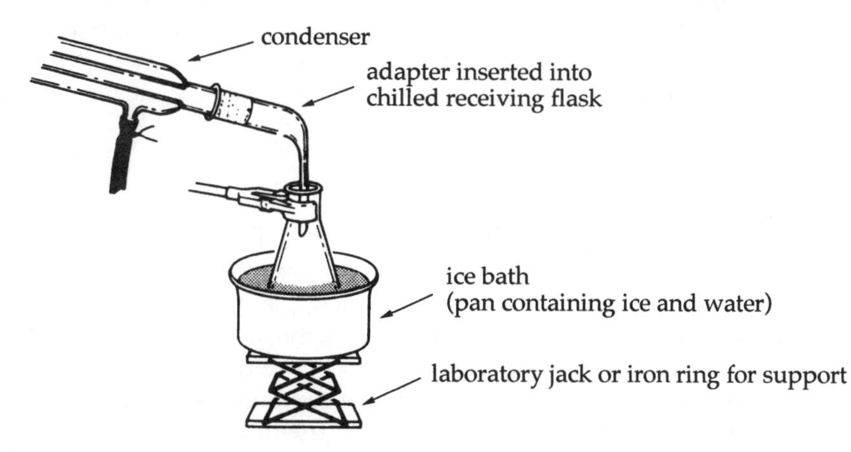

Figure 5.9 An ice-chilled distillation receiver.

(6) The Thermometer

Attach the thermometer last (and remove it first), because thermometers are expensive and easily broken. The easiest type of thermometer to insert is one with a ground-glass joint that fits a joint at the top of the distillation head. Neoprene adapters are available for attaching ordinary thermometers. Alternatively, a short piece of rubber tubing used as a sleeve can be used to hold the thermometer in place. A one-hole rubber stopper is not recommended because the hot vapors and condensate of the distilling liquid may dissolve some of the rubber, which will discolor the distillate. When attaching the thermometer, be sure to place the bulb just below the level of the sidearm, as shown in Figure 5.7, so that the bulb will be completely immersed in vapors of the distillate.

(7) The Actual Distillation

Before proceeding, check the water flow through the condenser and make sure that all ground-glass joints are snug. Plug the heating mantle into a rheostat; then plug the rheostat into the wall socket.

Slowly heat the mixture in the distilling flask to a gentle boil. You will then see the **reflux level** (the ring of condensate, or upper level of vapor condensing and running back into the flask) rise up the walls of the flask to the thermometer and sidearm. At this time, the temperature reading on the thermometer will rise rapidly until it registers the *initial boiling point,* which should be recorded. The vapors and condensate will pass through the sidearm and into the condenser, where most of the vapor will condense to liquid, and will finally drip from the adapter into the receiving flask.

The proper *rate of distillation* is one drop of distillate every 1–2 seconds. This rate is achieved by controlling the amount of heat supplied to the distillation flask. A slow rate means that not enough vapor is reaching the thermometer to give an accurate boiling point. A rapid rate results in uncondensed vapor being carried through the condenser and into the room. A rapid rate can also result in poor separation of components. It is generally necessary to increase the amount of heat applied to the distillation flask (by increasing the rheostat setting) during the course of a distillation. For distillations above 100°, insulate the head with crumpled aluminum foil to avoid temperature fluctuations.

(8) Collecting the Fractions

Volatile impurities are the first compounds to distill. This first fraction, called the **fore-run**, is generally collected separately. When the temperature has risen to the desired level and has been recorded, place a fresh receiver under the adapter to collect the main fraction. In some cases, the main fraction can be collected in a single receiver. In other cases, it should be collected as a series of smaller fractions. Each time you change a receiver, note the temperature reading and record the boiling range of the fraction. After checking the purity of a group of fractions, you may want to combine some of these fractions later.

Impurities that are higher-boiling than the desired material are generally not distilled but are left in the distillation flask as the residue. If higher-boiling impurities are present in large quantities, the temperature may rise from the desired level as the impurities begin to distill. However, the temperature frequently *drops* after the main fractions have distilled. This happens because not enough vapor and condensate are present in the head to keep the thermometer bulb hot. If the temperature drops at the end of a distillation, the last temperature to record is the highest temperature, before the drop occurred.

At the conclusion of a distillation, remove the heat source. Turn off a heating mantle and lower it from the flask immediately. Allow the entire apparatus to cool before dismantling it.

5.3 Microscale and Semi-Microscale Distillation

A. Simple Distillation

Distillation of small amounts of liquid requires the use of small, specialized glassware; otherwise the holdup volume (see p. 94) is as large as the amount of sample being distilled. The traditional distillation set-up shown in Figure 5.7 can be modified to microscale by using glassware that essentially looks the same, but that is much smaller. In some student laboratories, glassware that looks quite different from the traditional distillation set-up is employed. Remember: The goal of any microscale distillation apparatus is to shorten the path of distillation, thus reducing loss of product to the inside surfaces of the glassware.

Different manufacturers produce microscale glassware; each manufacturer incorporates a slightly different style of glass-to-glass fittings. One common type employs 14/10 standard-taper joints and 5–10 mL round-bottom flasks or 1–5 mL conical vials. Some manufacturers cast into the glassware screw threads on the outside of the glassware. Threaded glassware allows for a plastic screw clamp to firmly join the pieces of glassware together. Another type of microscale glassware employs rubber fittings to join the pieces of glassware together.

Two examples of microscale distillation set-ups that resemble traditional set-ups are shown in Figure 5.10. Often a water-jacketed condenser is not necessary, as air alone is sufficient to cool and condense the vapors.

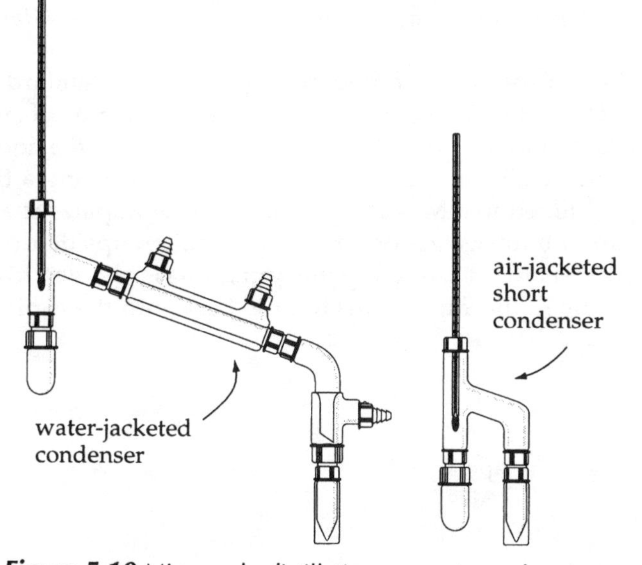
air-jacketed
short
condenser
water-jacketed
condenser

Figure 5.10 Microscale distillation apparatuses showing screw-type joints in a small version of a traditional distillation set-up.

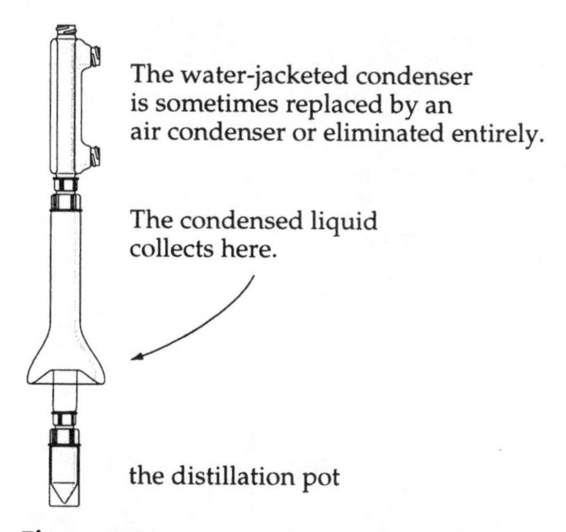
The water-jacketed condenser
is sometimes replaced by an
air condenser or eliminated entirely.
The condensed liquid
collects here.
the distillation pot

Figure 5.11 A microscale Hickman still. In this set-up, the distillate would be retrieved by reaching in through the top of the still head with a pipet. Some Hickman stills incorporate a side port for easy distillate retrieval.

Another effective set-up for microscale distillations incorporates a small *Hickman still* (Figure 5.11). In this type of apparatus, the volatile liquid distills from the pot and is collected on a small trough above the pot. A thermometer can be inserted through the top of the still head; however, the bulb of the thermometer must be entirely below the lip of the trough to obtain a reasonable boiling point.

B. Micro-boiling Points

If only a very small amount of material is available, the vapors do not sufficiently engulf the thermometer bulb, and an accurate boiling point cannot be obtained. In such instances, a **micro-boiling point** technique known as *Siwoloboff's method* is used.

A tiny drop of the liquid compound is placed in a standard, closed-end melting-point capillary tube. A second sealed-end capillary tube of smaller diameter is placed upside down in the first tube, with the open end in the liquid. (Drummond® micropipets work well for this purpose; seal one end using a flame.) The entire assembly is positioned in a Mel-Temp melting-point apparatus and heated until a constant stream of bubbles emerges from the smaller, upside-down capillary tube (see Figure 5.12). Then, the melting-point apparatus is allowed to cool. The temperature at which the liquid first begins to flow back into the capillary is the boiling point.

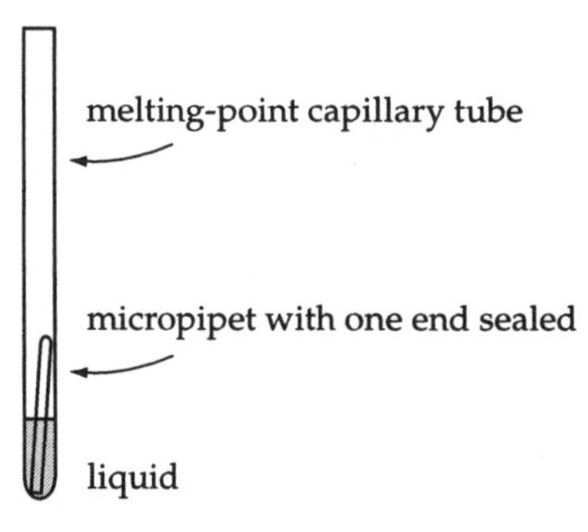

Figure 5.12 Capillary tube assembly for taking a micro-boiling point. The assembly fits into a Mel-Temp capillary melting-point apparatus (Figure 2.2, p. 42). (Note: this figure is greatly enlarged to show the detail of the micropipet in the liquid.)

Safety Notes

- Never add boiling chips to a hot liquid.
- Distillation of noxious or toxic substances should always be carried out in a fume hood.
- Never use a burner when distilling flammable substances.
- When distilling flammable substances, avoid allowing an excess of uncondensed vapors to flow into the room.
- When distilling at atmospheric pressure, always leave the apparatus open to the air at the adapter-receiver end. If you attempt a distillation with a closed system, the pressure buildup inside the apparatus may cause it to explode.
- Never carry out a distillation to dryness—always leave a small amount of residue in the distillation flask. The boiling residue will prevent the flask from overheating and breaking and will also prevent the formation of pyrolytic tars, which are difficult to wash out.
- Never dismantle a hot distillation apparatus—the hot residue or vapors might ignite when exposed to air.

Additional Resources

A list of suggested readings for the technique of distillation is published on the Brooks/Cole Web site, as well as additional problems. Please visit www.brookscole.com.

Problems

5.1 A mixture of two miscible liquids with widely different boiling points is distilled. The temperature of the distilling liquid is observed to plateau and then drop before rising again. Explain the temperature drop.

5.2 What will be the result of the following errors in a distillation process?
(a) The thermometer bulb is placed in the boiling liquid.
(b) The distillation flask is not securely attached to the distillation head, leaving an opening between the two pieces of glassware.
(c) Boiling chips are not added to the distillation flask.
(d) The water to the water-jacketed condenser is not turned on.

5.3 If a mixture is distilled rapidly, the separation of its components is poorer than if the mixture is distilled slowly. Explain.

5.4 From the graph in Figure 5.1, p. 78, estimate the following:
(a) the boiling point of acetone at 625 mm Hg
(b) the vapor pressure of acetone at room temperature (23°C)
(c) the percent of the total pressure contributed by acetone above an open beaker of this compound at room temperature (23°) and 760 mm

5.5 A 50% aqueous solution of ethanol (50 mL total) is distilled and collected in 10-mL fractions. Predict the boiling range of each fraction.

5.6 Calculate the mole fraction of each compound in the following mixtures.
(a) 95.0 g CH_3CH_2OH and 5.0 g H_2O
(b) 10.0 g CH_3OH, 10.0 g CH_3CH_2OH, and 10.0 g $CH_3CH_2CH_2OH$

5.7 Use the graph in Figure 5.5 to answer the following questions.
(a) Determine the composition of the liquid solution boiling at 100°.
(b) What is the composition of the vapor being given off?

5.8 A mixture of ideal miscible liquids C and D is distilled at 760 mm Hg pressure. At the start of the distillation, the mole percent of C in the mixture is 90.0, while that of D is 10.0. The vapor is condensed and found to contain 15.0 mole percent of C and 85.0 mole percent of D. Calculate the following.
(a) The partial vapor pressures (P) of C and D in this mixture
(b) The vapor pressures (P°) of pure C and D

5.9 Given the following mole fractions and vapor pressures for miscible liquids E and F, calculate the composition (in mole percent) of the vapor from a distilling ideal binary solution at 150° and 760 mm Hg. (The vapor pressures P_E° and P_F° are the values at 150°C.)

$X_E = 0.40$ $\quad$ $X_F = 0.60$
$P_E^{\circ} = 1710$ mm Hg $\quad$ $P_F^{\circ} = 127$ mm Hg

5.10 Given a mixture of compounds G and H. At 75°C,

$X_G = 0.50$ $\quad$ $X_H = 0.50$
$P_G^{\circ} = 900$ mm Hg $\quad$ $P_H^{\circ} = 400$ mm Hg

(a) Which compound has the lower boiling point?
(b) What is the partial vapor pressure of G? of H?
(c) If the applied pressure is 760 mm, will the mixture boil at 75°C?
(d) What is the composition of the vapor at 75°C?

Fractional Distillation

The technique of **fractional distillation** differs from simple distillation in only one respect: The vapor and condensate from the boiling liquid are passed through a **fractionation column** (Figure 6.1) before they reach the distillation head. This column contains a packing such as metal turnings or glass beads. As vapor rises through this column, it condenses on the packing and revaporizes continuously. Each revaporization of condensed liquid is equivalent to another simple distillation. Each of these "distillations" leads to a distillate successively richer in the lower-boiling component. Substantial enrichment of the vapor in the lower-boiling component occurs by the time the vapor reaches the head.

With a good fractionation column and proper operation, compounds with boiling points only a few degrees apart may be separated successfully. (Of course, an azeotropic mixture *cannot* be separated into its components by fractional distillation, because an azeotrope is a mixture of constant composition with a constant boiling point.)

6.1 Efficiency of the Fractionation Column

How well a fractionation column can separate a pair of compounds is called its *efficiency.* The efficiency of a column depends on both its length and its packing. In general, the greater the length of the column, the greater is its efficiency. For columns of the same length, an increase in the surface or an increase in the heat conductivity of the packing results in an increase in efficiency.

The efficiency of a column is reported in experimentally determined **theoretical plates**, where one theoretical plate is equivalent to one simple distillation. A distillation assembly with an efficiency of two theoretical plates is therefore equivalent to two simple distillations. The efficiency of a fractionating column is sometimes reported as HETP, which is ***H****eight of a column that is* ***E****quivalent to one* ***T****heoretical* ***P****late.* The lower the HETP value, the more efficient the column design. A distillation apparatus with one theoretical plate (a simple distillation head) separates compounds with a difference in boiling points of about 100° or more, but it does not give a good separation of compounds whose boiling points are closer together. At the other extreme, a column of 100 theoretical plates can separate a pair of compounds boiling as close as 2° apart. Most laboratory fractionation columns vary from 2 to about 15 theoretical plates. For example, a 25-cm column packed with glass beads or metal turnings has an efficiency of approximately 6 theoretical plates and, at best, can separate a binary mixture with a boiling-point difference of 30°–40°. Table 6.1

Table 6.1 Number of theoretical plates (TP) needed to separate a binary mixture.

Number of TP	*Approx. bp difference (°C)*
1	100
5	35
10	20
50	4
100	2

shows the number of theoretical plates needed to separate binary mixtures according to their boiling-point differences.

Fractional distillation must proceed at a slow rate so that vapor and liquid equilibrate at each "plate." This allows time for a part of the high-boiling component of the vapor mixture to condense and the same molar quantity of the lower-boiling component to vaporize at each plate. The fractional column must not be allowed to *flood* (show an excessive amount of liquid in one or more portions of the packing) during the distillation, because this in effect covers up the plates and decreases the opportunities for the vapor and the liquid to come into contact.

Other considerations in the efficiency of a fractional distillation set-up are the holdup and the reflux ratio. Holdup is the quantity of condensed vapor that remains on the packing when the distillation is stopped. If a very small amount of liquid is to be distilled, a column with a minimum holdup, even though it is less efficient, might be the column of choice. One such column, which contains no packing, is the Vigreux column.

The reflux ratio is the ratio of the amount of material that refluxes—condenses and flows back onto the column—to the amount of material that is removed through the sidearm as distillate. The reflux ratio should be at least 5, meaning that 5 times as much vapor condenses back into the system as flows out as distillate. A slow distillation has a high reflux ratio, while a fast one has a low ratio. ("Fast" and "slow" refer to the amount of distillate removed per unit of time.)

Table 6.2 Characteristics of some fractionation columns.*

Type of column	*Holdup (mL)*	*Theoretical plates (TP)*	*Height of each TP (HETP) (cm)*	*Separable bp difference (°C)*
Vigreux	1.5	3	8	50
glass helices	5	6	4	30
metal sponge	9	6	4	30
spinning band†	0.2	11–61	0.4–2	3–20

* Values are approximate due to individual column characteristics, compounds to be separated, operational differences, etc. These values are based on columns 10 mm in diameter, 25 cm long.

† A very efficient type of column used in research laboratories. The wide range of values arises from the different models of these columns.

6.2 Steps in Fractional Distillation

The following instructions for carrying out a fractional distillation assume that you will pack your own fractionation column. If a commercial distillation column is available, your instructor will provide directions for its use.

(1) Packing the Fractionation Column

The technique for packing a fractionation column depends on the packing material. Metal turnings or sponges are best pulled into the column with a wire hook. If the packing is glass beads, glass helices, or small metal turnings, first place a piece of metal sponge at the bottom of the column to support the packing (see Figure 6.1), then pour or drop in the pieces of packing. Regardless of the type of packing used, it should be loosely, but uniformly, packed. "Holes" in the packing will decrease efficiency, while spots of very tight packing may plug the column or cause it to flood (see below).

(2) Setting up the Apparatus

Assemble the apparatus shown in Figure 6.1, with the fractionation column clamped in a vertical position. View the column from more than one direction to make sure it is vertical. When distilling high-boiling compounds, insulate the column with glass cloth, dry rags, or a double layer of loosely wrapped aluminum foil. Whenever practical, however, leave the column uncovered so that you can observe the behavior of the liquid–vapor mixture in the column.

Clamp the distillation flask (1/2 to 2/3 full, containing boiling chips, and with its joint lightly greased) to the fractionation column. Clamp the receiving flask in position; then insert the thermometer into the distillation head.

(3) The Fractional Distillation

Heat the distillation flask slowly. When the solution boils, you will observe the ring of condensate rising up the fractionation column. If heating is too rapid and the condensate is pushed up too rapidly, equilibration between liquid and vapor will not occur and separation of the components will not be satisfactory.

If you heat the distillation flask too strongly before the column has been warmed by hot vapors and condensate, the column may *flood*, or show an excessive amount of liquid in one or more portions of the packing. Flooding is due to lack of equilibration between condensate and vapor and is more likely to occur if the packing has not been inserted uniformly. Ideally, the packing should appear wet throughout, but no portion of it should be clogged with liquid.

Flooding can be stopped by lowering the heat source. As the boiling of the liquid diminishes, the excess liquid in the column flows back into the distillation flask. At this time, resume heating, but more slowly than before. If flooding recurs, insulate the column as described in step (2) so that the vapors will have less tendency to condense. If the flooding is due to an incorrectly packed column, cool the apparatus, repack the column, and begin again.

(4) Collecting the Fractions

In a fractional distillation, read the boiling points and collect the fractions just as in a simple distillation. It is always better to collect a large number of small fractions than a few large ones. Small fractions of the same composition can always be combined, but a fraction that contains too many components must be redistilled.

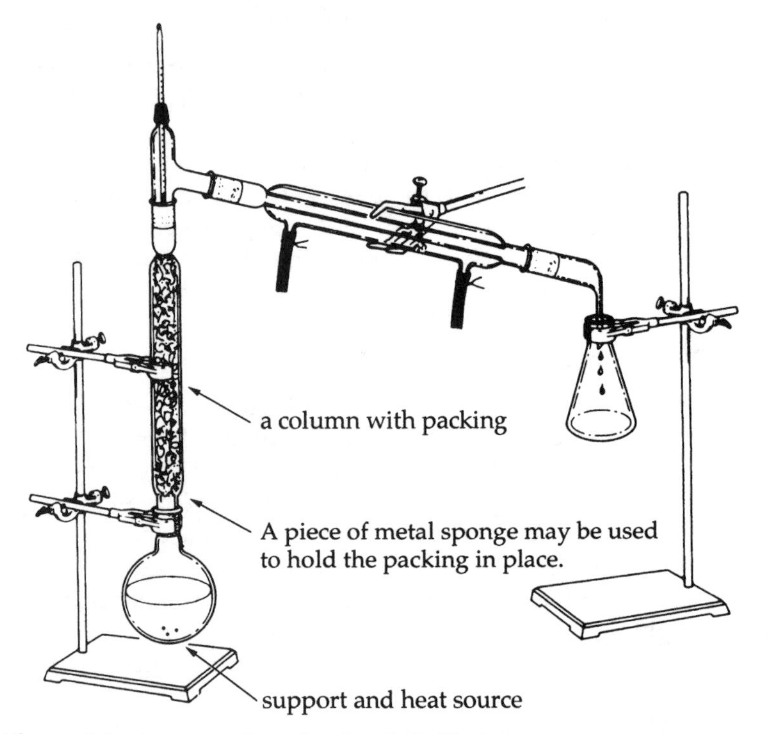
a column with packing
A piece of metal sponge may be used to hold the packing in place.
support and heat source

Figure 6.1 Apparatus for a fractional distillation.

> **Safety Note**
>
> - Before heating, check that all joints are snug, that fresh boiling chips have been added, and that the system is open to the atmosphere at the receiver.

6.3 Microscale Fractional Distillation

Microscale fractional distillations are performed with a smaller version of the fractionating column inserted in the microscale distillation apparatus above the distilling vessel. The microscale Y-adapter or a Hickman still is then placed above the fractionating column (refer to Figures 5.10 and 5.11). Stainless steel sponge is a convenient packing material for microscale fractionating columns because of its large surface area-to-size ratio. Spinning-band columns (Table 6.2) are also frequently used in microscale fractional distillations.

Additional Resources

A list of suggested readings for the technique of fractional distillation is published on the Brooks/Cole Web site, as well as additional problems. Please visit www.brookscole.com.

Problems

6.1 Draw an ideal distillation curve (temperature vs. milliliters of distillate) for a 50:50 mixture of two miscible liquids. How might one achieve such an ideal distillation?

6.2 Why is a fractionation column packed with glass helices more efficient than a Vigreux column of the same length and same diameter?

6.3 Which of the following circumstances might contribute to column flooding and why?
(a) holes in column packing
(b) packing too tight
(c) heating too rapidly
(d) column too cold

6.4 Explain why flooding in the fractionation column can lead to a poor separation of distilling components.

6.5 Referring to Tables 6.1 and 6.2, determine the approximate height of Vigreux columns needed to separate binary mixtures of compounds with the following boiling-point differences.
(a) 50°
(b) 25°
(c) 5°

6.6 Which of the following pairs of compounds be separated by simple distillation? Which would require fractional distillation? (Boiling points for organic compounds are found in both online databases and printed handbooks. See Technique 17, p. 198.)
(a) *n*-butyl acetate and methyl acetate
(b) 3,3-dimethyl-2-butanone and 1-chloro-2-propanone
(c) cyclopentanol and 2-methyl-1-propanol
(d) 2-heptanone and *n*-pentyl propionate

6.7 For each pair of compounds that require fractional distillation for separation in Problem 6.6, how many theoretical plates would be required? What type of fractionating column would give the proper number of theoretical plates?

6.8 A chemist has a small amount of a compound (bp 65°) that must be fractionally distilled, yet the chemist does not want to lose any of the compound to holdup on the column. What can the chemist do?

Vacuum Distillation

When the pressure over a liquid is reduced, the liquid boils at a lowered temperature. (Refer to Figure 5.1, p. 78; the boiling-point–pressure diagram for acetone.) As its name implies, **vacuum distillation** (simple or fractional) is distillation under reduced pressure. Because the reduction in pressure lowers the boiling point, vacuum distillation is used for distilling high-boiling or heat-sensitive compounds.

7.1 Boiling Point and Pressure

At pressures near atmospheric pressure, a drop in pressure of 10 mm Hg generally lowers the boiling point of a substance by about 0.5°.* At the low pressures usually used in vacuum distillation, halving the pressure reduces the boiling point about 10°. For example, a compound with a boiling point of 100° at 20 mm Hg would boil at about 90° at 10 mm Hg.

Figure 7.6 (p. 105) shows a diagram, called a *nomograph*, for estimating boiling points at various pressures. To use the nomograph, follow these steps.

(1) Assume that the boiling point of a compound at one particular pressure is known. Find the reported boiling point on scale A and the pressure (P_1) on scale C. Connect these two points with a transparent ruler.

(2) Read the boiling point at atmospheric pressure (760 mm Hg) at the intercept on scale B.

(3) To find the boiling point at a different pressure, P_2, connect the normal atmospheric boiling point on scale B to P_2 on scale C. Read the new boiling point from scale A.

EXAMPLE A compound boils at 80° at 1.0 mm Hg. What is its boiling point at 20 mm Hg?

(1) Connect 80° on scale A to 1.0 mm on scale C.

(2) Read the atmospheric boiling point on scale B; it is 250°.

---

* Another way to estimate the corrected boiling point is expressed in the following formula: $T_{corr} = T_{obs} + 0.00010(760 - P)(T_{obs} + 273)$, where T_{corr} is the corrected boiling point, T_{obs} is the observed boiling point, and P is the barometric pressure. This estimation is useful for locations that are significantly higher in altitude than sea level.

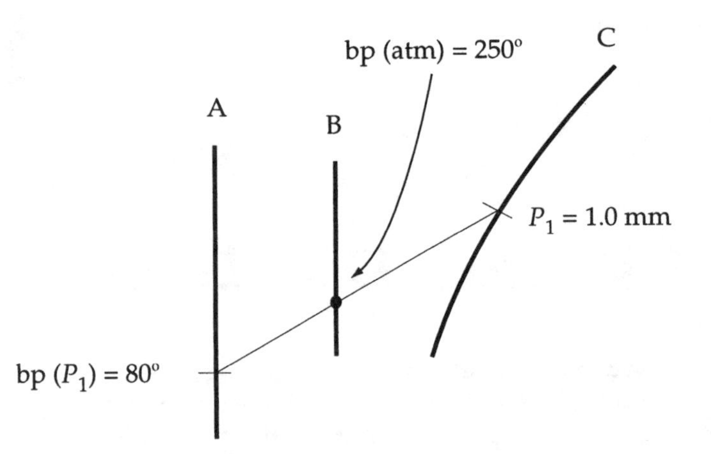

(3) Connect 250° on scale B to 20 mm on scale C. Read the boiling point at 20 mm Hg from scale A. In our example, the estimated boiling point on scale A is 130°.

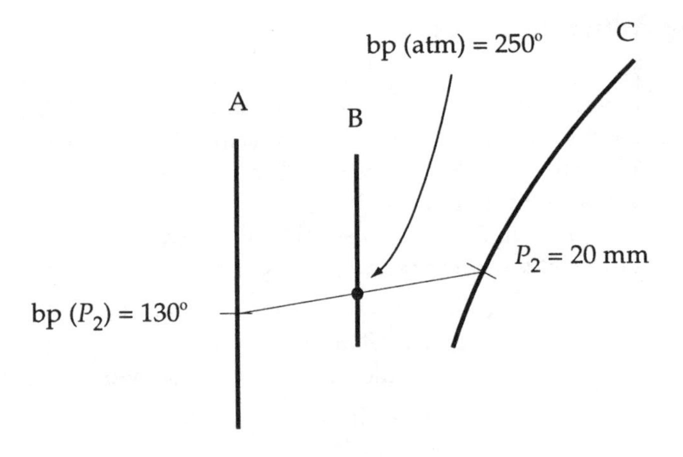

7.2 Apparatus for Simple Vacuum Distillation

Figure 7.1 shows a typical student vacuum distillation apparatus. If the apparatus used in your laboratory is substantially different (for instance, if you are using microscale glassware), your instructor will tell you how to assemble it.

The principal difference between a vacuum and an atmospheric distillation set-up is that in vacuum distillation, the system must be airtight and able to hold a vacuum. A vacuum adapter rather than a bent adapter must be used, since this adapter has a standard-taper joint connection for a round-bottom flask and a tubing connection for the vacuum source. The glassware must be able to withstand the vacuum; for example, large, thin-walled Erlenmeyer flasks cannot be used.

Under vacuum, most solvents boil well below room temperature; therefore, any solvent in the material to be distilled must be removed prior to distillation.

Use vacuum grease to seal all glass-to-glass connections. Stopcock grease flows too easily under heat to hold a vacuum. Use only heavy-walled vacuum tubing. Tighten all rubber-tubing connections (including a rubber thermometer

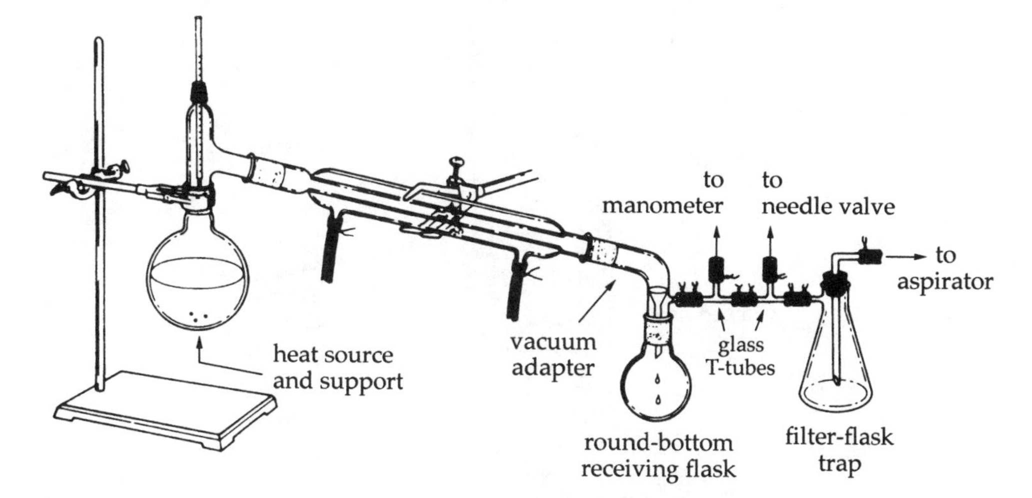

Figure 7.1 A simple vacuum distillation assembly. Each flask and the condenser are clamped in place (not all clamps are shown). All connecting heavy-walled rubber tubing is wired.

adapter) by wrapping a wire around the rubber and twisting it with pliers. Assemble the apparatus completely and test for air leaks by applying the vacuum before placing any liquid in the distillation flask.

Because ordinary boiling chips do not function well in a vacuum distillation, special boiling chips with smaller pores must be used to prevent bumping. Alternatively, a very fine capillary may be used to introduce a thin stream of air or N_2 bubbles to the boiling liquid (see Figure 7.2). CAUTION: Do not pass air through oxygen-sensitive compounds.

Vacuum fraction collectors. If a vacuum adapter and round-bottom flask are used, the fore-run cannot be separated from the main fraction without your stopping the distillation and disassembling the apparatus. Therefore, if more than one fraction is to be collected, a vacuum fraction collector should be used. Two types of vacuum fraction collectors are shown in Figure 7.3.

Vacuum sources. A vacuum can be obtained by using either a mechanical pump or a water aspirator. Mechanical pumps must be used for distillations at pressures lower than 20 mm Hg. If a mechanical pump is used, it must be protected by traps to prevent vapors of distillate from passing into it. Your instructor can show you how to use the pump. For pressures of 20 mm Hg and above, a water aspirator is usually satisfactory.

One principal disadvantage of an aspirator results from fluctuating water pressure. If the water pressure drops, the vacuum in the distillation apparatus can suck water from the aspirator into the receiving flask. To ensure that such an event does not occur, position a water trap between the aspirator and the distillation apparatus (see Figure 7.1). This trap is similar to the trap for vacuum filtration, except that a one-hole stopper is used.

Pressure regulation and measurement. Place a needle valve (or other pressure regulation device) and a manometer (pressure-measuring device) between the vacuum source and the vacuum adapter. The needle valve allows you to adjust the

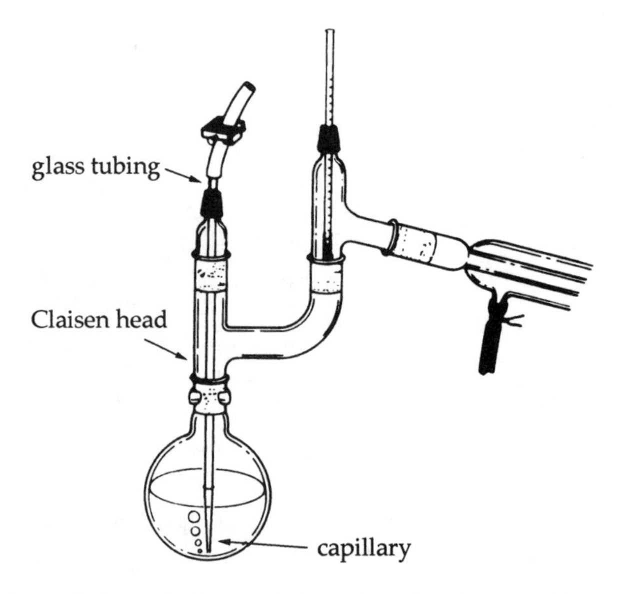

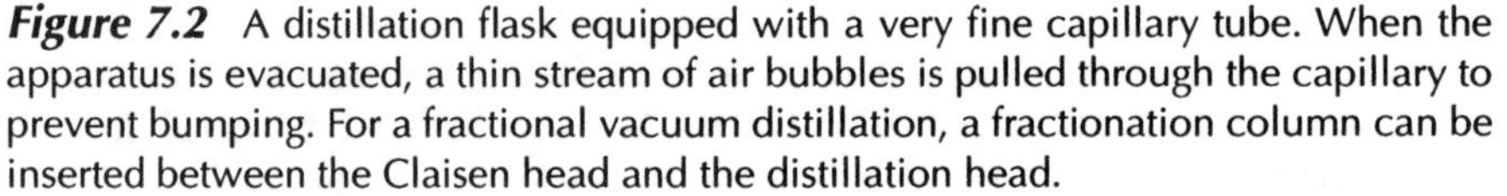
Figure 7.2 A distillation flask equipped with a very fine capillary tube. When the apparatus is evacuated, a thin stream of air bubbles is pulled through the capillary to prevent bumping. For a fractional vacuum distillation, a fractionation column can be inserted between the Claisen head and the distillation head.

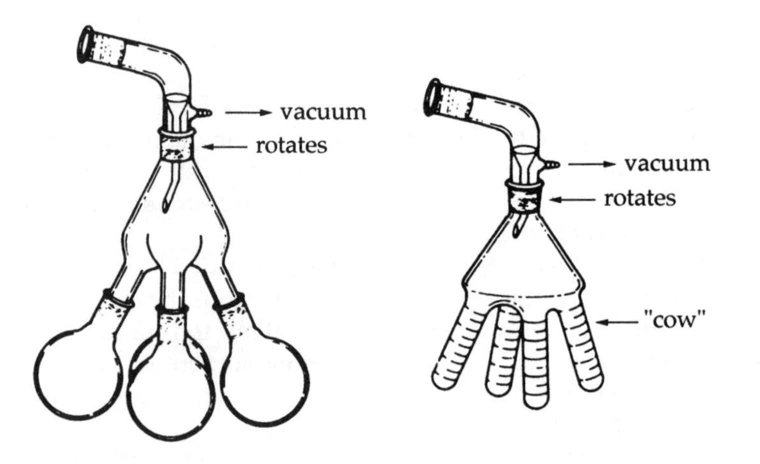

Figure 7.3 Vacuum adapters and receivers that allow the collection of more than one fraction during the course of a distillation.

pressure to the desired level by bleeding air into the system. If a needle valve is not available, use the base of a Fischer or Meker burner. Connect the vacuum tubing to the gas inlet and adjust the vacuum with the valve at the base of the burner normally used to adjust the height of the flame. Pressure control with a needle valve or burner is satisfactory in distillations conducted at a pressure greater than 20 mm Hg.

Closed-end manometers, as shown in Figure 7.4, are suitable for measuring pressures above 20–25 mm Hg. In either of these manometers, the difference in the levels of mercury, in mm, is equal to the pressure in the system. For distillations at pressures lower than 20 mm, another pressure-measuring device, such as a "tilting McLeod gauge" must be used (see Figure 7.5). The technique for using a McLeod gauge is outlined in Figure 7.5.

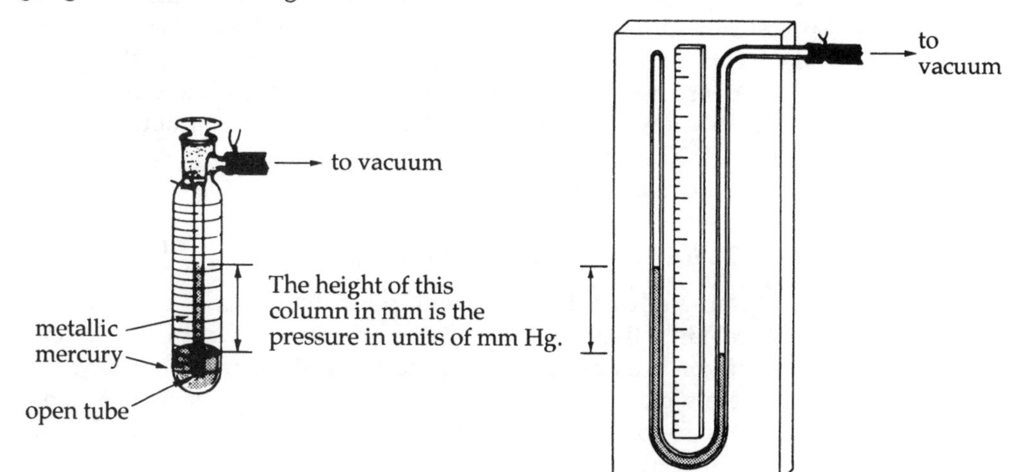

Figure 7.4 Typical closed-tube manometers.

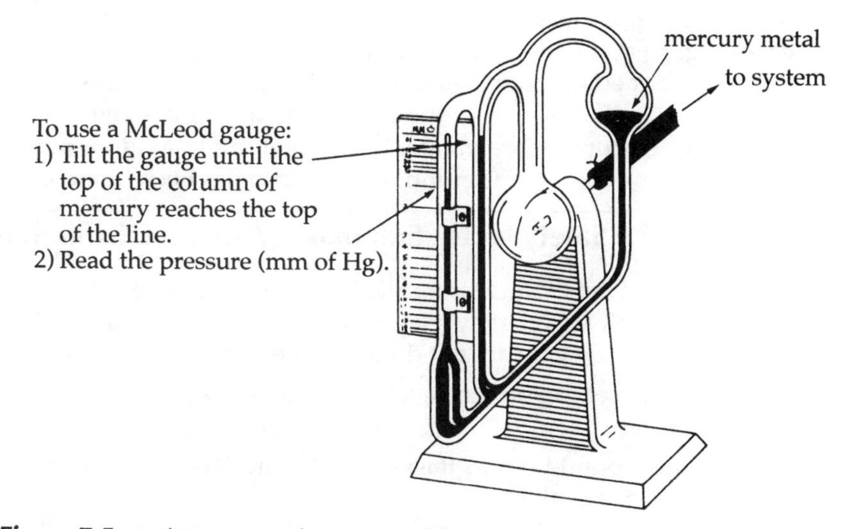

Figure 7.5 A tilting McLeod gauge, used for measuring very low pressures (0.01–10 mm Hg). The gauge does not automatically record changes in pressure but must be swiveled each time a pressure reading is made.

7.3 Steps in Vacuum Distillation

(1) Test the Vacuum with Apparatus Empty

Tare a receiving flask and assemble a simple vacuum distillation apparatus (Figure 7.1) as described in the discussion. With an empty distillation flask, test the seals by applying vacuum with the needle valve closed. You should be able to obtain a pressure of approximately 20 mm Hg using an aspirator or a much lower pressure with a mechanical pump. If you cannot obtain this low pressure, use the pinch clamp to isolate various parts of the system to find the leak(s).

After checking the empty system, break the vacuum by opening the needle valve slowly while watching the mercury in the manometer rise. (If the vacuum is broken very rapidly, the rising mercury can have sufficient momentum to break the glass in the manometer.)

(2) Add the Mixture to be Distilled

Using a funnel with a stem, pour the material to be distilled (cool and solvent-free) into the distillation flask and add a few carbon or Micro-Porous® boiling chips if you are not using a capillary. Apply a full vacuum with the needle valve completely closed. You should be able to attain nearly the same low pressure as when the distillation flask was empty. If not, a new leak has developed; seal it before proceeding.

(3) Set the Pressure and Distill

Adjust the needle valve on the system so that the pressure is at the desired level; then let the system stand for a few minutes until the pressure is no longer changing. Heat the mixture to a boil with a heating mantle, and distill. (If the distillation is stopped and then resumed, add fresh boiling chips to the mixture.) At the end of the distillation, lower the heating mantle and allow the system to approach room temperature before breaking the vacuum, to avoid igniting the hot residue. Do not turn off the aspirator until the vacuum has been broken.

Safety Notes Summary: Vacuum Distillation

- Whenever vacuum is applied to a closed system, there is the danger of an implosion. Safety glasses must be worn by all students in the lab, even those who are not directly involved in the vacuum distillation.
- Before setting up the vacuum distillation apparatus, check all glassware for stars or cracks.
- Only round-bottom flasks or pressure flasks should be used in a vacuum distillation.
- Do not pass air through oxygen-sensitive compounds.

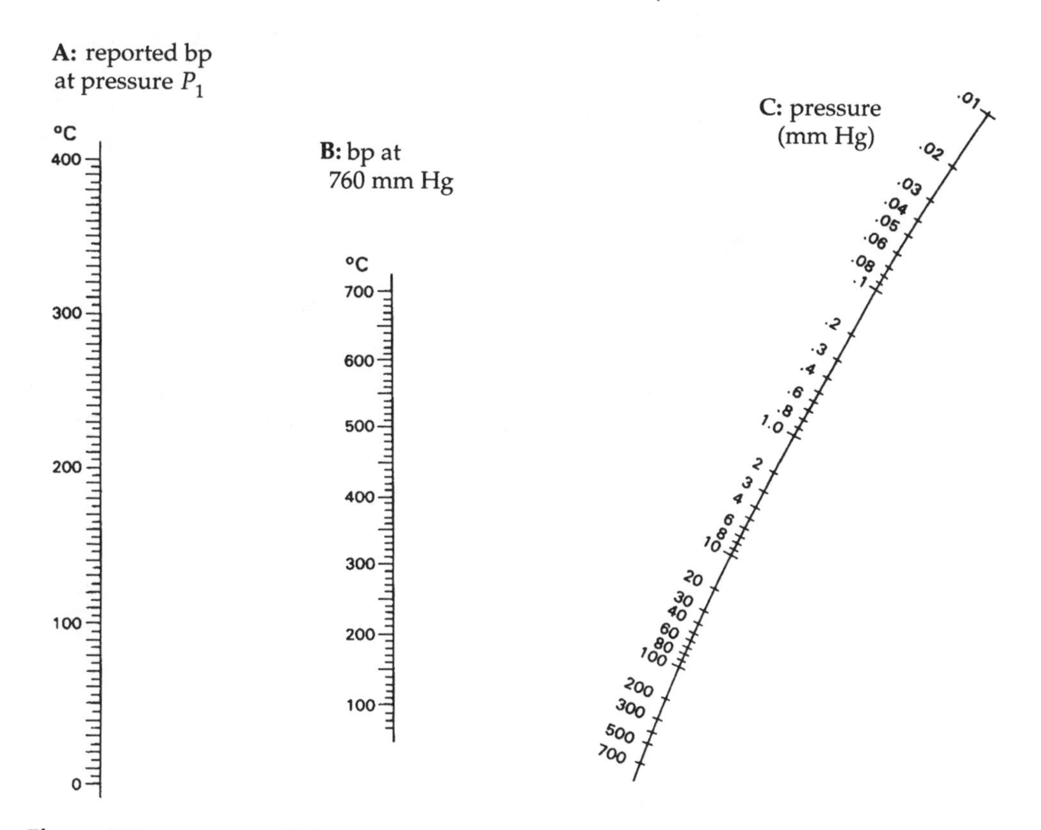

Figure 7.6 A nomograph for estimating boiling points at different pressures.

Additional Resources

A list of suggested readings for the technique of vacuum distillation is published on the Brooks/Cole Web site, as well as additional problems. Please visit www.brooks-cole.com.

Problems

7.1 Predict the boiling points of the following compounds at the pressure indicated.
(a) water at 745 mm Hg
(b) dichloromethane (bp 40° at 760 mm) at 737 mm
(c) benzene (bp 80° at 760 mm) at 765 mm

7.2 Using the nomograph in Figure 7.6, approximate the boiling points of the following substances at atmospheric pressure (760 mm Hg).
(a) rotenone, bp 210° (0.5 mm Hg)
(b) 1,6-hexanediol, bp 132° (9 mm Hg)

7.3 Approximate the boiling points of the compounds in problem 7.2 at 25 mm Hg.

7.4 At low pressures such as used in vacuum distillation, reducing the pressure by half reduces a boiling point by about 10°. If a compound boils at 180° at 10 mm Hg, what would be its approximate boiling point at the following pressures?

(a) 15 mm Hg
(b) 2.0 mm Hg

7.5 Explain why a vacuum distillation apparatus should be checked for leaks before the material to be distilled is placed in the flask.

7.6 Explain how you would purify octyl acetate from a mixture of octyl acetate and methylene chloride. (You will first need to look up the boiling point of octyl acetate in a suitable handbook, Internet site, or online source. See Technique 17, p. 198.)

Steam Distillation

Steam distillation is the distillation of a mixture of water (steam) and an organic compound or a mixture of organic compounds. The organic compound must be insoluble (immiscible) in water for the steam distillation to be successful. Immiscible mixtures such as water and an organic compound do not behave like solutions. The components of an immiscible mixture boil at *lower temperatures than the boiling points of any of the components*. Thus, a mixture of high-boiling organic compounds and water can be distilled at a temperature less than 100°, the boiling point of water. Natural products, such as flavorings and perfumes found in leaves and flowers, are sometimes separated from their sources by steam distillation. This technique can sometimes replace vacuum distillation and other separation techniques in the organic laboratory.

8.1 Characteristics of Steam Distillation

The principle that gives rise to steam distillation is that the total vapor pressure of a mixture of *immiscible* liquids is equal to the *sum of the vapor pressures of the pure individual components*. The total vapor pressure of the mixture thus equals atmospheric pressure—and the mixture boils—at a lower temperature than the boiling point of any one of the components singly. For a steam distillation to be successful, the component to be isolated must have a vapor pressure higher than those of the other organic components in the mixture.

This boiling behavior of an immiscible mixture is different from that of miscible liquids, where the total vapor pressure is the sum of the *partial* vapor pressures of the components. Therefore, with immiscible mixtures, the mole fraction of the components in the mixture is not important.

For immiscible liquids:

vapor pressures of pure components

$$P_{total} = P^{\circ}_{A} + P^{\circ}_{B}$$

For miscible liquids:

$$P_{total} = \underbrace{X_A P^{\circ}_{A}} + \underbrace{X_B P^{\circ}_{B}}$$

partial vapor pressures

In a distillation of two immiscible liquids, the number of moles of each component in the vapor (and thus in the distillate) is proportional to the vapor pressure of the pure component.

For the steam distillate:

$$\frac{\text{moles of A}}{\text{moles of B}} = \frac{P^{\circ}_{A}}{P^{\circ}_{B}}$$

8.2 Amount of Water Needed in a Steam Distillation

The previous equations allow us to calculate the actual amount of water needed to steam-distill a particular compound. Substituting the definition of moles and rearranging the equation, we see that the weight ratio of the two components is dependent only on the molecular weights and vapor pressures.

$$\frac{wt_A/MW_A}{wt_B/MW_B} = \frac{P^{\circ}_{A}}{P^{\circ}_{B}}$$

$$\frac{wt_A}{wt_B} = \frac{P^{\circ}_{A} \times MW_A}{P^{\circ}_{B} \times MW_B}$$

In order to calculate how much water is needed, we need only the vapor pressure of water at the distillation temperature, the vapor pressure of the compound to be distilled at that temperature, and their molecular weights. The vapor pressure of water at various temperatures is well known. A few of these values are listed in Table 8.1. (The *Handbook of Chemistry and Physics* contains a more extensive table.)

In general, the vapor pressure of the compound will not be known, but it can be estimated using the vapor pressure of water and the total atmospheric pressure.

vapor pressure of water at the distillation T

$$P^{\circ}_{B} = P_{total} - P^{\circ}_{A}$$

If the structure of the organic compound is known, then its molecular weight can be calculated. The molecular weight of water, of course, is well known.

Table 8.1 Vapor pressure of water at various temperatures.

Temperature (°C)	*Vapor pressure (mm Hg)*
60	149
70	234
80	355
90	526
95	634
100	760

EXAMPLE At 760 mm Hg, bromobenzene (C_6H_5Br, bp 155°) steam-distills at approximately 95°. Calculate the approximate amount of water needed to steam-distill 20 g of bromobenzene.

$$\frac{wt_{H_2O}}{wt_{C_6H_5Br}} = \frac{P^{\circ}_{H_2O} \times MW_{H_2O}}{P^{\circ}_{C_6H_5Br} \times MW_{C_6H_5Br}}$$

$$= \frac{634 \times 18}{(760 - 634) \times 157}$$

$$= \frac{0.58 \text{ g water}}{\text{g bromobenzene}}$$

Thus, for 20 g of bromobenzene, we would need to distill 20 x 0.58 g, or about 12 g, of water.

8.3 Apparatus for Steam Distillation

The simplest way to perform a steam distillation is to add water to the material to be distilled and distill in the ordinary manner. However, if a large volume of water is needed, the water may boil away before the steam distillation is completed. Fig-

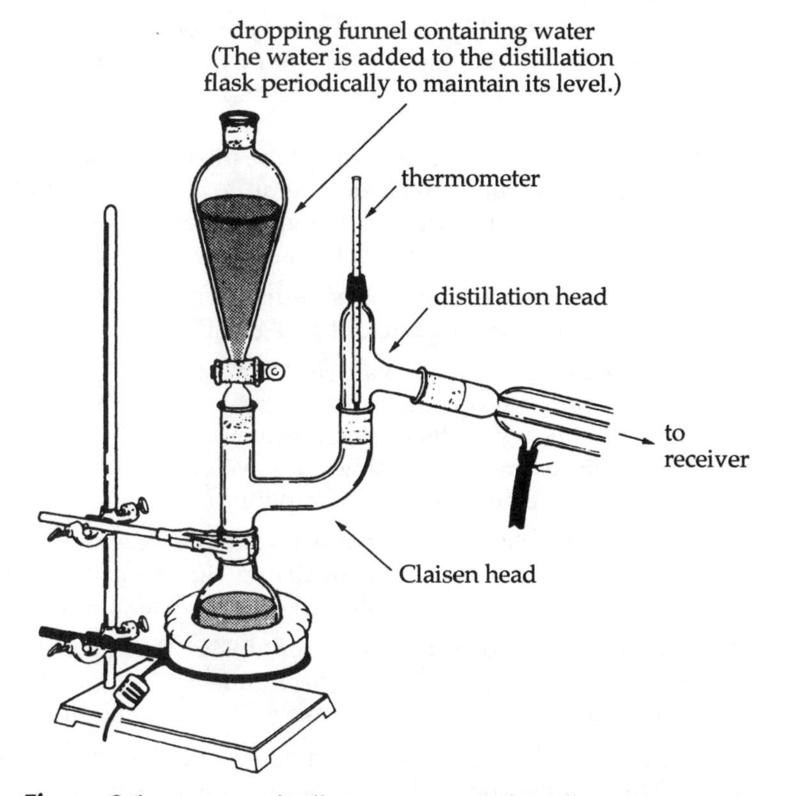

Figure 8.1 A steam distillation apparatus that allows water to be added to the distillation flask during the distillation.

ure 8.1 shows a set-up that allows the water in the distillation flask to be replenished.

Live steam can also be used in a steam distillation through the use of a steam line. One advantage of using live steam is that the mixture need not always be heated to boiling, even though some heating is necessary to prevent excessive condensation of the steam in the flask.

The quantity of distillate collected in a steam distillation depends on the vapor pressure of the organic component. A compound with a high vapor pressure will steam-distill with a small amount of water. If the vapor pressure of the organic material to be isolated is low, a greater amount of distillate must be collected. The steam distillation is complete when only pure water distills. This may be determined by noting when the distillate is clear and no longer composed of two phases.

Additional Resources

A list of suggested readings for the technique of steam distillation is published on the Brooks/Cole Web site, as well as additional problems. Please visit www.brooks-cole.com.

Problems

8.1 A mixture of immiscible liquids (both water-insoluble) is subjected to steam distillation. At 90°, the vapor pressure of pure water is 526 mm Hg. If the vapor pressure of compound A is 127 mm Hg and that of B is 246 mm Hg at 90°:

(a) what is the total vapor pressure of the mixture at 90°?
(b) would this mixture boil at a temperature above or below 90°?
(c) what would be the effect on the vapor pressure and boiling temperature by doubling the amount of water used?

8.2 Suggest a reason why benzene steam-distills at the rate of 1 g/0.1 g H_2O, but nitrobenzene distills at the rate of 1 g/4 g H_2O.

8.3 At 99°, the vapor pressure of water is 733 mm Hg.

(a) At standard atmospheric pressure, what is the vapor pressure of a compound being steam-distilled at this temperature?
(b) If the compound has a molecular weight of 180, how much water is required to distill 1.0 g of the compound?

8.4 At 99.6°, water has a vapor pressure of 750 mm Hg and quinoline (immiscible with water) has a vapor pressure of 10 mm Hg. What weight of water must be distilled for each gram of quinoline in a steam distillation at 760 mm Hg?

8.5 Aniline (MW 93) co-distills with water at 98.2°. The partial vapor pressure of water at 98.2° is 712 mm Hg. You need to isolate 10 g of aniline. What is the minimum amount of water it will take to steam-distill this amount of aniline?

8.6 Limonene (MW 136) is a pleasant-smelling liquid found in lemon and orange peels. The bp of limonene is 175°, but it co-distills with water at 97.5°. If the vapor pressure of water at 97.5° is 690 mm Hg, what percentage of the steam-distillate is limonene?

Sublimation

Sublimation is a process whereby a solid is purified by vaporizing and condensing it without its going through an intermediate liquid state.

Solid compounds that evaporate (that is, pass directly from the solid phase to the gaseous phase) are rather rare; solid CO_2 (dry ice) is a familiar example of such a compound. Even though both solids and liquids have vapor pressures at any given temperature, most solids have very low vapor pressures. In order for a solid to evaporate, it must have an unusually high vapor pressure compared with those of other solids. For a solid compound to exhibit such a high vapor pressure, it must have relatively weak intermolecular attractions. One factor that contributes to weak intermolecular attractions is the shape of the molecules. Many compounds that evaporate readily contain molecules that are roughly spherical or cylindrical shapes that do not lend themselves to strong intermolecular attractions. Table 9.1 lists some solids that can be sublimated in the laboratory.

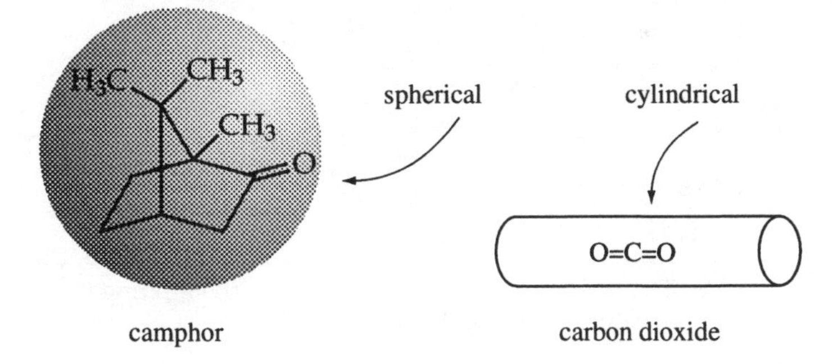

Sublimation can be used to purify some solids, just as distillation can be used to purify liquids. In sublimation, nonvolatile solid impurities remain behind when the sample evaporates, and condensation of the vapor yields the pure solid compound. Sublimation has the advantages of being reasonably fast and clean because no solvent is used. Unfortunately, most solid compounds have vapor pressures too low for purification in this fashion. Also, sublimation is successful only if the impurities have much lower vapor pressures than that of the substance being purified. It would be practical, for example, to purify technical-grade iodine, which is contaminated with inorganic salts, by sublimation. It would not be practical, however, to separate camphor from isoborneol (from which commercial camphor is synthesized) because *both* compounds readily sublime.

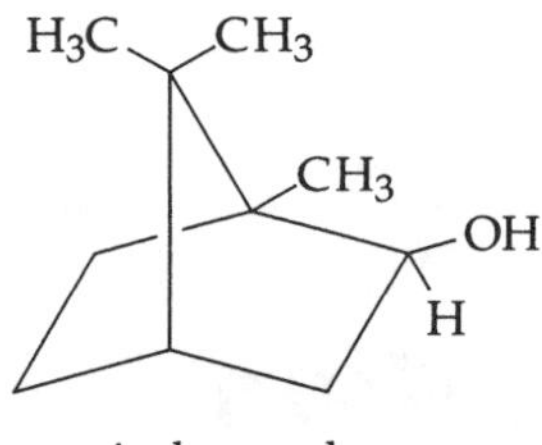

isoborneol

The vapor pressure of a solid increases with temperature, just as it does for liquids. Therefore, evaporation can be facilitated by heating the solid, but not to its melting point. The rate of evaporation can also be increased by subliming the solid under a vacuum; however, a very efficient cooling surface must be used for the condensation so that the solid's vapor is not lost into the vacuum system.

A simple sublimation apparatus for a student laboratory is shown in Figure 9.1. Figure 9.2 shows an apparatus for carrying out a sublimation under vacuum.

Table 9.1 Vapor pressures of some solids at their melting points.

Compound	*mp (°C)*	*Vapor pressure (mm Hg) at the melting point**
hexachloroethane	186	780
camphor	179	370
iodine	114	90
hydroquinone	169	14.1
p-dichlorobenzene	53	8.5
naphthalene	80	7
benzoic acid	122	6

* The melting point of a compound is the limiting temperature for its sublimation.

9.1 Atmospheric Sublimation

The atmospheric sublimation apparatus, shown in Figure 9.1, can be used only for a solid with a relatively high vapor pressure. Place the sample in a filter flask equipped with a water-cooled cold finger or a test tube filled with ice. Warm the flask on a hot plate or in a water bath, taking care not to melt the solid. Crystal growth on the test tube (and on the cooler flask sides) soon occurs. Periodically, cool the apparatus, remove the cold finger or test tube carefully, and scrape the sublimed crystals into a suitable tared container. Determine the melting point of the sample with a sealed capillary.

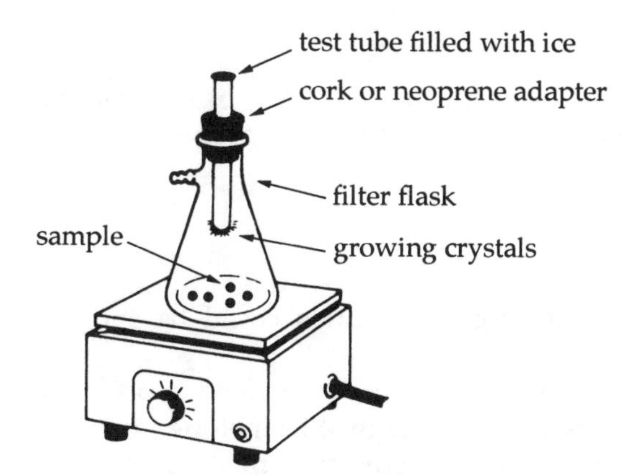

Figure 9.1 A sublimation apparatus using a filter flask, ice-filled test tube, and hot plate.

9.2 Vacuum Sublimation

For a vacuum sublimation, choose a sidearm test tube to hold the sample, and insert a smaller test tube fitted into a large-holed rubber stopper (see Figure 9.2). Place ice in the inner tube, and connect the sidearm, using heavy-walled rubber tubing, to a trap (preferably chilled in an ice bath) and then to an aspirator or other vacuum source. (See Technique 1, p. 30, for an illustration of the trap and a discussion of using the aspirator.)

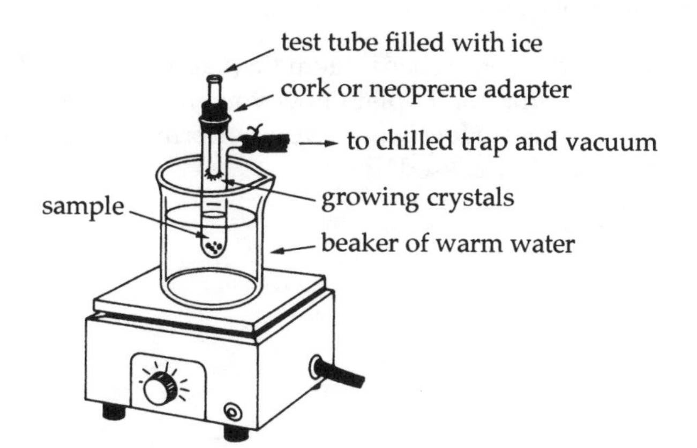

Figure 9.2 A vacuum sublimation apparatus using a large sidearm test tube.

Additional Resources

A list of suggested readings for the technique of sublimation is published on the Brooks/Cole Web site, as well as additional problems. Please visit www.brooks-cole.com.

Problems

9.1 Which of the following compounds could be subjected to sublimation at atmospheric pressure?
(a) Compound A: vapor pressure at its melting point = 770 mm Hg
(b) Compound B: vapor pressure at its melting point = 400 mm Hg
(c) Compound C: vapor pressure at its melting point = 10 mm Hg

9.2 In the preceding problem, which compounds could be vacuum sublimated?

9.3 Which of the following compounds would be likely to evaporate readily? Explain your answer.
(a) $CH_3(CH_2)_{14}CH_3$
hexadecane

(b) 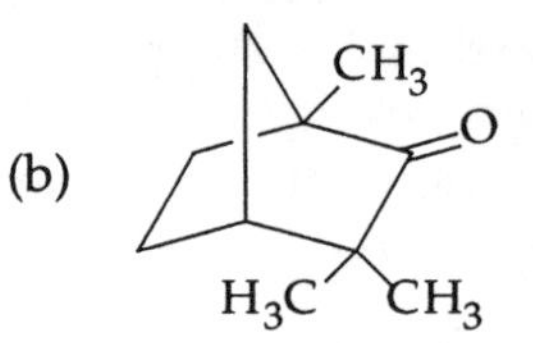

fenchone (in oil of fennel)

(c) 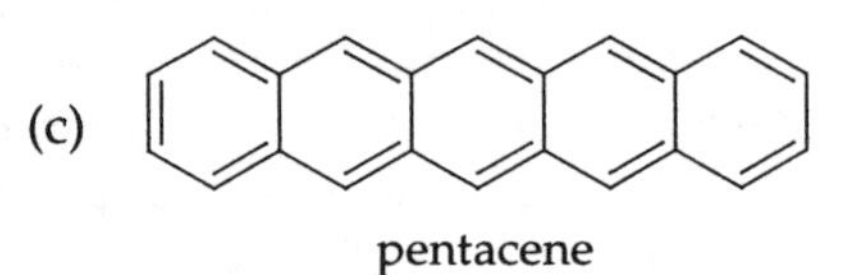
pentacene

9.4 You have a mixture of *p*-nitrobenzaldehyde (vapor pressure 0.01 at its melting point of 106°) and camphor. Explain how you would separate these two compounds by sublimation. How would you perform the sublimation if benzoic acid instead of camphor is used?

Technique 10

Refractive Index

A refractive index is a physical property, like a boiling point, that can be used as one piece of evidence in determining the identity and purity of a liquid compound. **Refraction** is the bending of a ray of light as it passes obliquely from one medium to another of different density. Refraction arises from the fact that light travels more slowly through a denser substance than it does through one that is less dense. In organic chemistry, we are interested in the refraction of light as it passes through a liquid layer. Refraction is useful because the degree of refraction depends on the structure of the compound.

The refractive index (n) of a particular substance is defined as the ratio of the speed of light in a vacuum or in air (the values are very close) to the speed of light in the substance.

$$\text{refractive index, } n = \frac{\text{speed of light in air}}{\text{speed of light in the substance}}$$

The refractive index is measured with an instrument called a **refractometer**, which determines the angle of refraction of light between the liquid and a prism. A typical refractometer is shown in Figure 10.1.

Different wavelengths of light are refracted different amounts. This is the reason that sunlight can be split into the visible spectrum (a rainbow) by droplets of water. When the refractive index is used as a physical constant, only one wavelength of light is used, usually the sodium D line, at 589.3 nm. This single wavelength can be obtained from either a sodium lamp or an ordinary white light with a prism system.

Besides being dependent on wavelength, the refractive index is temperature-dependent. For this reason, the temperature is always specified when a refractive index is reported. A typical refractive index for a compound, such as for benzene or ethanol, is reported as in the following examples:

benzene, n_D^{20} 1.5011 ethanol, n_D^{20} 1.3611

where D refers to the wavelength (the sodium D line) and the superscript number is °C

The refractive index is a very sensitive physical property. Unless a compound is extremely pure, it is almost impossible to duplicate a literature value exactly. For example, a sample of benzene from a fractional distillation might exhibit n_D^{20} 1.4990 instead of the reported value of n_D^{20} 1.5011. However, the closer the observed refractive index is to the reported value, the more pure the compound is likely to be.

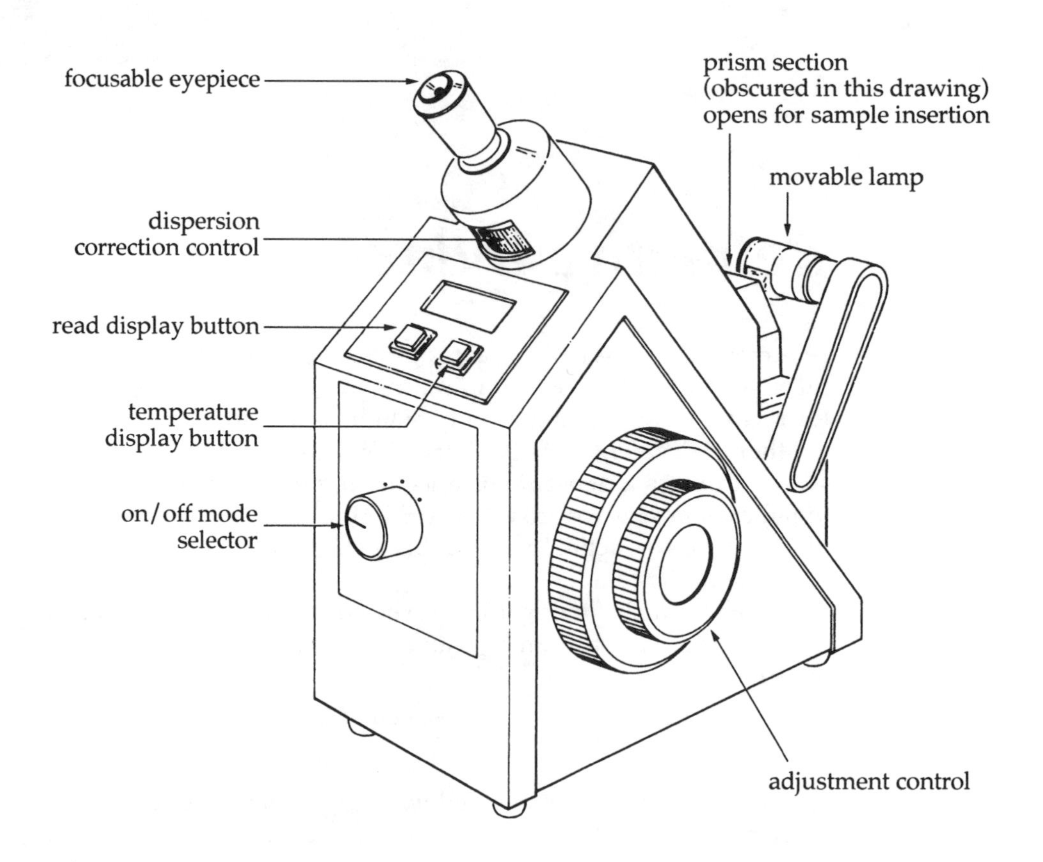

Figure 10.1 A typical refractometer (Source: Courtesy of Leica Inc., Buffalo, New York.)

In terms of structure, the refractive index is a function of the polarizability of the atoms and groups within molecules. A more polarizable molecule exhibits a higher refractive index. Thus, we find that alkyl iodides (with large polarizable iodo-substituents) and most aromatic compounds (with polarizable pi systems) have high refractive indexes, generally greater than 1.5000. By contrast, most aliphatic compounds exhibit refractive indexes between 1.3500 and 1.5000.

10.1 Correcting for Temperature Differences

The refractive index varies inversely with temperature. An increase in temperature causes a liquid to become less dense and almost always causes a decrease in refractive index. It has been determined experimentally that the average temperature correction factor for a wide variety of compounds is 0.00045 unit for each degree Celsius. The following examples show how to adjust a refractive index to a different temperature.

EXAMPLE 1 (adjustment to a higher temperature)

To adjust n_D^{18} 1.4370 to n_D^{20}, proceed as shown.

Calculate factor:

2°C x 0.00045/°C = 0.00090

Subtract from measured value:

$$n_D^{20} = 1.4370 - 0.0009 = 1.4361$$

EXAMPLE 2 (adjustment to a lower temperature)

To adjust n_D^{23} 1.5066 to n_D^{20}, proceed as shown.

Calculate factor:

3°C x 0.00045/°C = 0.00135 (round to 0.0014)

Add to measured value:

$$n_D^{20} = 1.5066 + 0.0014 = 1.5080$$

10.2 Steps in Using a Refractometer

Your instructor will give you specific instructions on the use of the refractometer in your laboratory. The following general steps, however, apply to most refractometers.

(1) Prepare the Prism Section

Open the prism section, which can be done mechanically. If the prism surfaces are not absolutely clean, clean them with a few drops of ethanol and a lens-cleaning tissue. (CAUTION: The prism surfaces are very easily scratched. Do not touch these surfaces with anything hard, such as an eye dropper, the end of a glass rod, or a metal spatula. Clean them gently by dabbing with the tissue.)

(2) Place the Liquid on the Prism Surface

Place 1–3 drops of the liquid to be analyzed on the lower prism surface; then close the prism section. The prism surfaces fit closely together; when the section is closed, the liquid is forced to spread out as a thin film. Or close the prism section and then apply a drop or two of liquid to the edge of the gap between the prism surfaces. A thin film will form between the surfaces by capillary action.

(3) Position the Light

Position the lamp to shine directly into the glass prism of the instrument.

(4) Adjust the Instrument

Look through the eyepiece and use the appropriate knobs to focus the cross hairs and to bring a light area and a dark area into view. The appearance of the field is illustrated in Figure 10.2. If the dividing line between the light and dark areas is fringed with red or blue, the prism system is not directing the sodium D line exactly through the sample. The prisms may be adjusted by a dispersion correction knob. The field is in proper adjustment when there is a sharp dividing line, with no color, between the light and dark areas.

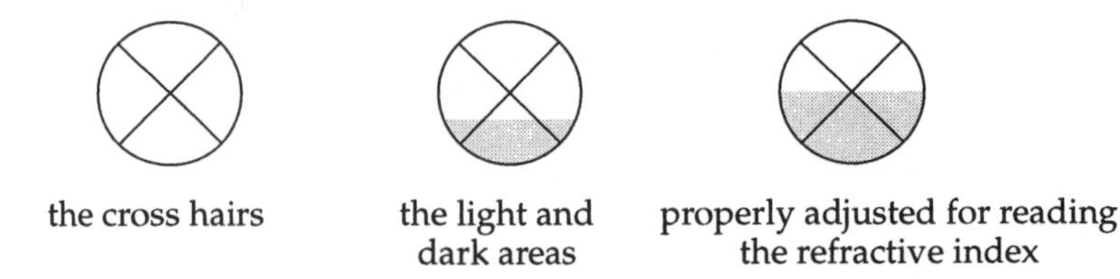

Figure 10.2 The field of view through the eyepiece of a refractometer.

(5) Center the Dividing Line

To determine the refractive index, move the dividing line between the light and dark areas exactly to the center of the cross hairs, read the value for the refractive index, and record the temperature.

(6) Clean the Instrument

Clean the prism surfaces with a few drops of ethanol and a lens-cleaning tissue. (Please note the caution in step 1.)

Problems

10.1 Correct the following refractive indexes to 20°C.

(a) n_D^{22} 1.4398

(b) n_D^{30} 1.4702

(c) n_D^{16} 1.3962

(d) n_D^{18} 1.4022

10.2 Compound A has a refractive index of 1.4577 at 20°, while compound B (miscible with compound A) has a refractive index of 1.5000 at the same temperature.

(a) How could you experimentally determine if the refractive indexes of mixtures of A and B follow a linear relationship to the mole percent compositions of the mixtures?

(b) If the relationship is linear, what is the composition (in mole percent) of a mixture that has a refractive index of 1.4678 at 20°C?

10.3 When measuring a refractive index, what should you do if

(a) no dividing line between light and dark areas can be observed through the eyepiece?

(b) the dividing line can be seen, but it is fringed with color?

Column Chromatography

The term *chromatography* was coined by the Russian botanist Mikhail Tswett in the late nineteenth century. Tswett studied plant pigments and found that he could separate green chlorophylls and orange carotenes from green leaf extracts using a narrow glass tube filled with calcium carbonate. In his words: "like light rays in the spectrum, the different components of a pigment mixture, obeying a law, separate on the calcium carbonate column". Tswett called this separation process the *chromatographic method*, from the Greek *chromato*, meaning color, and *graphy*, which refers to writing, or, "written in color." The use of chromatography progressed slowly until the 1930s and the post–war era. Today, chromatography methods are the most modern and sophisticated means of separating the components of a mixture of compounds.

In chromatography, a mixture is separated by distributing the components between a **stationary phase** and a **moving phase**. This rather broad definition covers the techniques of thin-layer chromatography, gas chromatography, paper chromatography, ion-exchange chromatography, high-pressure liquid chromatography, size exclusion chromatography, and column chromatography. In each of these techniques, the mixture to be separated is placed on the stationary phase (a solid or a liquid), and the mobile phase (a gas or a liquid) is allowed to pass through the system. The different compounds in the mixture move through the adsorbent at different rates because of physical differences (such as size or vapor pressure) and because of different interactions (adsorptivities, solubilities, and so on) with the stationary phase. Thus, the individual compounds in a sample become separated from one another as they pass through the adsorbent and can be either collected or detected, depending on the chromatographic technique and the quantity of sample used.

The use of chromatographic techniques is not limited to organic chemistry—these techniques find wide use in a variety of scientific areas. For example, gas–liquid chromatography is used in criminology laboratories for blood-alcohol tests; thin-layer chromatography and gas–liquid chromatography are used in environmental and biology laboratories; and all types of chromatography are used in medical laboratories for both research and routine analyses. A relatively new tool, high-performance liquid chromatography (HPLC), combines many of the best features of two chromatographic techniques: gas–liquid chromatography and column chromatography.

Column chromatography is used to separate, and thus purify, the components of a mixture. The adsorbent is contained in a column constructed of glass (or sometimes metal). A mixture of compounds is applied to the top of the column and then solvent is added. If the solvent is allowed to flow down the column by gravity, or percolation, it is called **gravity column chromatography**. If the solvent is forced down the column by positive air pressure, it is called **flash chromatography**. Both methods are used routinely in modern organic chemistry laboratories to analyze and separate mixtures of compounds.

11.1 Gravity Column Chromatography

In **gravity** column chromatography, a vertical glass column is packed with a polar adsorbent along with a solvent. The sample is added to the top of the column; then additional solvent is passed through the column to wash the components of the sample, one by one (ideally), down through the adsorbent to the outlet. Figure 11.1 illustrates the column and the technique.

The sample on the column is subjected to two opposing forces: the solvent dissolving it and the adsorbent adsorbing it. The dissolving and the adsorption constitute an equilibrium process, with some sample molecules being adsorbed and others leaving the adsorbent to be moved along with the solvent, only to be readsorbed farther down the column. A compound (usually a nonpolar one) that is very soluble in the solvent, but not strongly adsorbed, moves through the column relatively rapidly. On the other hand, a compound (usually a more polar compound) that is attracted to the adsorbent moves through the column more slowly.

Because of the differences in the rates at which compounds move through the column of adsorbent, a mixture of compounds is separated into *bands*, each compound forming its own band that moves through the column at its own rate. Figure 11.1 shows the formation of a pair of bands (which are not always visible). The bands are finally washed out of, or *eluted* from, the bottom of the column, each to be collected in a separate flask. Table 11.1 lists the usual order of elution of different types of compounds.

A. The Adsorbent

The selective action of an adsorbent is very similar to the action of decolorizing charcoal, which selectively adsorbs colored compounds. In fact, activated carbon is sometimes used as an adsorbent in column chromatography. The adsorption process is due to intermolecular attractions, such as dipole–dipole attractions or hydrogen bonding.

$$\overset{\delta-}{X}-\overset{\delta+}{R} \quad \cdots \quad \delta{+}Al_2O_3\,\delta{-} \qquad\qquad RO-H \cdots Al_2O_3$$

dipole–dipole attractions — a hydrogen bond

Different adsorbents attract different types of molecules. A highly polar adsorbent strongly adsorbs polar molecules but has little attraction for nonpolar molecules such as hydrocarbons. This is why nonpolar compounds are usually eluted first and more polar compounds are eluted later. Because adsorbents differ in their adsorbing power, an adsorbent chosen for a particular chromatographic separation depends in part on the types of compounds being separated. Table 11.2 lists a few of the common adsorbents used to pack chromatography columns.

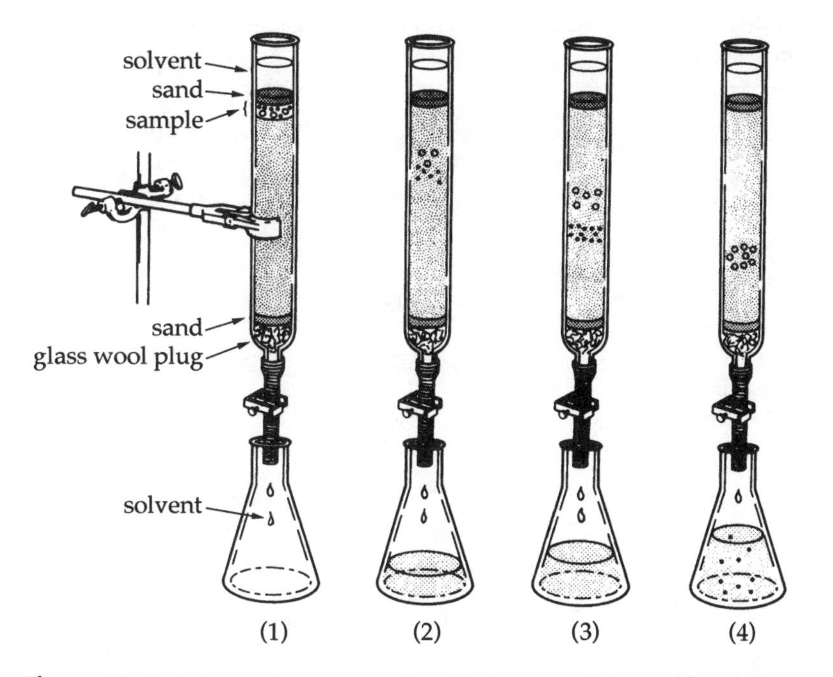

Figure 11.1 Column chromatography. (1) The sample has just started to move into the column of adsorbent. (2) and (3) As more solvent is passed through the column, the sample moves down and begins to separate into its components because of differences in attraction to the adsorbent and solvent. (4) The faster-moving compound is eluted into a flask.

Silica gel (SiO_2, silicic acid) and alumina (Al_2O_3) are common column chromatography adsorbents. Both of these adsorbents are polar, with silica gel being the less polar of the two. Alumina is quite sensitive to the amount of water bound to it: The higher its water content, the less polar sites it has to bind organic compounds, and thus the less "sticky" it is. This adsorptive power or *activity* of alumina is designated as I, II, or III, with I being the most active. Alumina is usually purchased as Activity I and deactivated with water before use according to specific procedures. Alumina also comes in three forms: acidic, neutral, and basic. Different forms of alumina are chosen according to the acid–base character of the compounds being separated.

Column chromatography adsorbents are sold in different *mesh* sizes, as indicated by a number on the bottle label: "silica gel 60" or "silica gel 230–400" are two examples. The number refers to the mesh size of the sieve used to size the silica—it is the number of holes in the mesh or screen through which the crude silica particle mixture is passed in the manufacturing process. If there are more holes per unit area, those holes are smaller, thus allowing only smaller silica particles go through the sieve. Therefore, the larger the mesh number, the smaller the adsorbent particles.

Adsorbent particle size affects how the solvent flows through the column. Smaller particles (higher mesh values) are used for flash chromatography; larger particles (lower mesh values) are used for gravity chromatography. For example,

Table 11.1 The expected elution order of organic classes in column and thin-layer chromatography.

	Name of class	*General formula*
increasing polarity ↓	alkanes	RH
	alkenes	$R_2C{=}CR_2$
	ethers	R_2O
	halogenated hydrocarbons	RX
	aromatic hydrocarbons	benzene, 2-methylnaphthalene (CH_3) (etc.)
	aldehydes and ketones	$RCH{=}O$, $RC({=}O)R$
	esters	RCO_2R
	alcohols	ROH
	amines	RNH_2, R_2NH, R_3N
slow	carboxylic acids	RCO_2H

230–400 mesh silica gel is used for flash columns and 70–230 mesh is used for gravity columns.

Before you begin a column chromatography experiment, check the ability of a particular adsorbent to separate a particular sample using various solvents and thin-layer chromatography (TLC; Technique 12). TLC will also indicate the minimum number of compounds in the sample.

The quantity of adsorbent needed depends on the sample, the solvent, and the column. Generally, a minimum of 20–30 g of adsorbent per gram of sample is necessary. This is packed into a column either as a slurry or dry. The column height should be at least ten times its diameter. For the separation of about 0.1–2 g of sample, a column with a diameter of about 3 cm and a height of 30–100 cm is convenient.

Table 11.2 Typical adsorbents used in column chromatography.

increasing adsorptive power for polar molecules ↑	alumina
	activated charcoal
	magnesium silicate (Florisil)
	silica gel
	inorganic carbonates
	starch, cellulose, sucrose

B. The Solvent

Generally, solvents are organic compounds. They, too, can be adsorbed by the packing in a chromatography column, and they compete with the sample for positions on the adsorbent. However, nonpolar solvents are not as highly attracted to an adsorbent as are other organic compounds.

The action of a solvent, or a series of solvents, can be used to effect a separation. Assume that you have a mixture of polar and nonpolar compounds and wish to separate the compounds by column chromatography. After adding the sample to the top of the column, you would begin by dripping a nonpolar hydrocarbon solvent like hexanes or petroleum ether through the column. These solvent molecules are not adsorbed to any degree. Polar molecules in the sample are more attracted to the adsorbent than to the nonpolar solvent; therefore, the polar molecules are selectively held back on the column. The nonpolar components of the mixture are not strongly adsorbed and are highly soluble in the nonpolar solvent. These compounds move down the column, at a relatively rapid rate, to be eluted and collected.

To remove the more polar compounds from the column, you would use a more polar solvent. If you were using a hexanes as the solvent, you would add to it a small amount of a solvent such as ethyl acetate or diethyl ether. Then, you would gradually increase the percent of the more polar solvent so that the polarity of the eluting solvent system would be gradually increased. Table 11.3 lists some common solvents in order of their polarity. Note the similarities between this list and the list in Table 11.1.

The polarity of the eluting solvent should not be changed rapidly, especially when low-boiling solvents are used. The heat generated by interactions of solvents and adsorbents can cause a low-boiling solvent to boil in the column and thus ruin the homogeneity of the column.

Table 11.3 Some solvents used in column chromatography.

	Name	*Structure*
increasing polarity ↓	alkanes (hexanes, petroleum ether)	—*
	toluene	$C_6H_5CH_3$
	halogenated hydrocarbons (methylene chloride, chloroform)	CH_2Cl_2, $CHCl_3$
	diethyl ether	$(CH_3CH_2)_2O$
	ethyl acetate	$CH_3CO_2CH_2CH_3$
	acetone	$(CH_3)_2C{=}O$
	alcohols	CH_3OH, CH_3CH_2OH
	acetic acid	CH_3CO_2H

* Petroleum ether and hexanes are mixtures of alkanes; see Table 1.1.

C. Collecting the Fractions

In the introductory organic laboratory, chromatographic separations are often carried out with mixtures of colored compounds so that their separation on the column can be observed. When one colored component is about to be eluted from the column, a fresh flask should be placed under the column to collect it.

In actual practice, most chromatographic separations involve colorless compounds, whose presence must be detected in other ways. Typically, a series of same-size fractions (5-mL or 25-mL, for example) are collected in tared flasks. It is better to collect many small fractions than a few large ones. These fractions are evaporated and weighed, or they are concentrated and tested by thin-layer chromatography (Technique 12). If TLC is used for monitoring the fractions while the separation is in progress, the operator can check the purity of the compounds as they are eluted and can estimate the number of compounds remaining on the column at any time. Alternatively, GC or HPLC can be used to monitor the progress of the separation.

D. Steps in Gravity Column Chromatography

There are many ways to prepare a column for and carry out a column chromatography; therefore, it is difficult to present a general procedure. We will present only one technique: the preparation of a silica gel column. Even if your instructor would like you to use a different procedure, we suggest you read through the following steps to learn some of the important features of column packing and use.*

(1) Preparing a Silica Gel Column

A column must be packed uniformly and contain no holes or air bubbles. The following procedure for a 50- to 75-mL column is one way to prepare a homogeneous wet column. For difficult separations, the silica gel should be sieved just prior to column preparation.†

Mount a clean chromatography column (3 cm x 38 cm) as shown in Figure 11.1. Check the column from two directions to make sure it is vertical. With the stopcock (or clamp) closed, pour 50 mL of *n*-hexane (or other nonpolar solvent) into the column. Then, using a glass rod or glass tubing, push a very small plug of glass wool to the bottom of the column. Use the rod to push out any air entrapped in the glass wool. Then add enough sand to the column to cover the glass wool. Using solvent from a dropper, wash down any sand clinging to the sides of the column. Level the sand by tapping the glass column with your finger. Finally, drain about 20 mL of the hexane from the column to ensure that all the air bubbles have been displaced.

Weigh 20 g of sieved silica gel into a beaker and add 50 mL of hexane. Mix the slurry with a spatula. When completely mixed, the slurry should appear almost translucent and should contain no white lumps of silica gel or air bubbles. Adjust

---

* The procedure outlined here was developed for a separation of 0.10 g of a mixture of the colored compounds *o*- and *p*-nitroaniline, 5% of each in 95% ethanol. (The ethanol was used for ease in column loading.) One additional milliliter of ethanol was used to wash the solution onto the column packing.

† The final particle size should pass an 80-mesh screen but be retained on a 115-mesh screen (Tyler Standard Screen Scale).

the stopcock on the column to allow about 5 mL/minute of solvent to pass into the receiving flask. Remix the slurry and pour it in a continuous stream into the chromatography column. Wash the beaker immediately with about 20 mL of hexane and add this wash to the top of the column.

While the hexane is still draining through the column, wash any silica gel on the sides of the column onto the packing, using hexane from a dropper. Allow the hexane to drain from the column until only about 1 cm remains on top of the packing; then close the stopcock. As the hexane drains from the base of the column, the silica gel settles and forms the solid support for the chromatographic separation. The packing is very delicate and will be useless if it is allowed to drain dry of solvent.

(2) Loading the Column

Allow the hexane to drain from the column until the top of the packing is just free of liquid. Using a dropper, transfer 1 mL of a solution of the mixture to be separated directly onto the top of the packing. Do not let the solution run down the sides of the glass onto the packing. Open the stopcock to allow this solution to flow into the packing. Stop when the top of the packing is just free of liquid.

Measure out no more than 1 mL of solvent and, using a dropper, wash the solution, along with any splatters on the sides of the column, into the packing, again stopping when the top of the packing is just free of the liquid.

Using a clean dropper, add about 3 mL of hexane and drain it into the packing as before, being careful not to disturb the top of the packing. Repeat the process with a second 3-mL portion of hexane. Finally, using a dropper, gently add 10 mL of hexane to the top of the packing; then very carefully pour 90 mL of hexane down the side of the column onto the packing.

(3) Developing the Column

Adjust the flow rate of the hexane from the column to 1–1.5 mL per minute. (A faster rate of flow will cause the packing to form channels, so that the compounds will "tail," or not move down evenly.) Continue at this rate until about 1 cm of hexane remains on top of the packing.

Carefully add 100 mL of the 3% ethyl acetate–hexane eluting solvent (or other mixture) to the top of the column and continue the eluting process. If the compounds are visible on the column, collect each compound as it elutes from the column in a tared flask. When only one compound remains on the column, a polar solvent can be used to wash it off quickly. If the compounds are not visible on the column, collect 25-mL volumes in tared flasks.

In either case, evaporate the solvent from the flasks. Weigh the flasks to determine the amount of material collected. Thin-layer chromatography (Technique 12) is an excellent analytical tool for analyzing the purity of the compounds you isolate from column chromatography.

Traditional gravity column chromatography, as described here, suffers from a number of practical disadvantages. Primarily, it is time-consuming and tedious; it also requires very large amounts of solvents. Traditional column chromatographic techniques are therefore being replaced by **flash chromatography** and related instrumental techniques called **high-performance liquid chromatography (HPLC)**.

11.2 Flash Chromatography (Macroscale)

Classic column chromatography described in the previous section tends to be time-consuming and sometimes results in poor separation because the compound bands have a tendency to tail, especially when the quantities of material being separated exceed 1–2 g. In an effort to overcome these problems, a method known as **flash chromatography** has largely replaced classical column chromatography. Moderate resolution can be achieved with samples weighing 0.01–10.0 g. The total time required for column packing, sample application, and elution can be as little as 10–15 minutes.

A. Apparatus

The apparatus required consists of a chromatography column that has a Teflon stopcock and a standard-taper glass joint at the top, as shown in Figure 11.2. The diameter of the glass column depends on the sample size, as listed in Table 11.4. A flow controller, consisting of a needle valve and a connector for the gas, is attached with a standard taper joint to the top of the column. The rate of elution is controlled by air or nitrogen pressure (about 20 psi) applied to the top of the column. The adsorbent used for flash chromatography has a smaller particle size (230–400 mesh) than that used for classical column chromatography.

B. Steps in Flash Chromatography

(1) Select a Solvent System

Select a solvent system that gives a good separation between the desired compound and any impurities and that moves the desired component to an $R_f = 0.35$ on a silica gel TLC plate (R_f is defined in Technique 12, p. 133). Solvents found to be

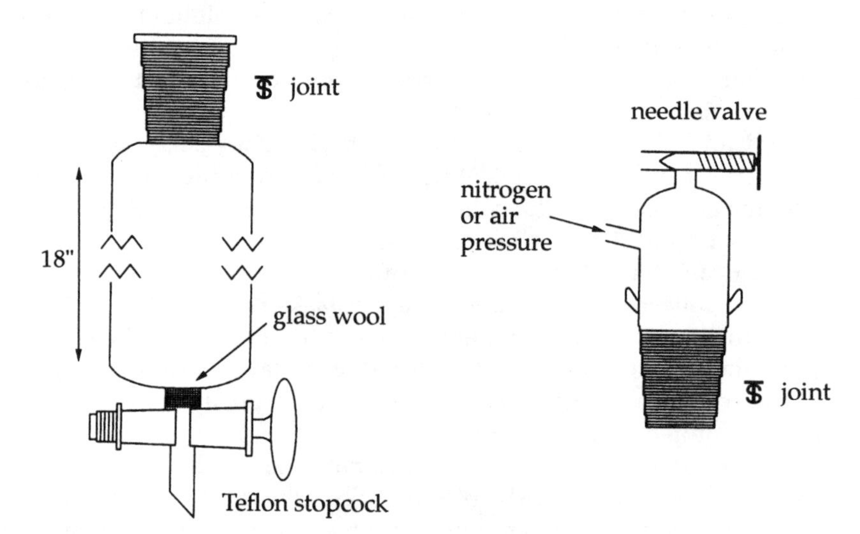

Figure 11.2 Apparatus for flash chromatography. (From *Theory and Practice in the Organic Lab* by Landgrebe, **1993**, Wadsworth, Inc. Reprinted by permission.)

especially useful include ethyl acetate–petroleum ether (bp 30°–60°) for general separations and either acetone or methylene chloride with petroleum ether (bp 30°–60°) for separations of polar compounds.

(2) Prepare the Column

Select a column of the appropriate diameter (see Table 11.4), put a plug of glass wool at the bottom of the column, add a thin layer of 50–100-mesh dry sand as a platform, and (with the stopcock open) fill it with 5–6 in. of dry 40–60 μm silica gel. Tap the column gently; then add a $\frac{1}{8}$-in. layer of dry sand on top of the silica gel.

Carefully fill the column with solvent, attach and secure the flow controller, and, with the bleed valve open, turn on a small flow of air or nitrogen. Place your finger over the bleed port so that the pressure builds up rapidly in the column and compresses the silica gel. This procedure forces the air pockets out of the bottom of the column. Maintain the pressure until all the air has been expelled; otherwise, the column may fragment and be ruined when the pressure is released.

Release the pressure and readjust the needle valve to maintain a slight pressure so that excess solvent is forced through the column, but do not let the top of the adsorbent go dry. The solvent used in packing the column can be reused to elute the sample.

Table 11.4 Choosing a column and estimating eluting volume for flash chromatography.*

Column diameter, mm	*Volume of eluant, mL*	*Amount of sample that can be loaded, mg*†	*Typical fraction size, mL*
10	100	40–100	5
20	200	16–400	10
30	400	360–900	20
40	600	600–1600	30
50	1000	1000–2500	50

* Reprinted with permission from W. C. Still, M. Kahn, and A. Mitra, *J. Org. Chem.* **1978**, *43*, 2923. Copyright 1978 American Chemical Society.

† The minimum value is suitable if the difference in R_f (ΔR_f) of the compounds to be separated is ≥0.1; the maximum value corresponds to ΔR_f of ≥0.2. R_f is defined on p. 133.

(3) Load the Column

Dissolve the sample as a 20–25% solution dissolved in the eluting solvent. Use a Pasteur pipet to add this solution to the top of the column. Apply a slight pressure to force the sample onto the adsorbent. If the sample is not very soluble in the elution solvent, apply it to the column with a solvent containing a little more of the polar component of the eluting solvent.

(4) Develop the Column

Apply adequate pressure to the column to achieve a drop in solvent level of about 2 in./minute. Collect fractions until all of the recommended solvent has been used (Table 11.4).

Because the time required to elute the column is very short (5–10 minutes), it is best to collect the eluent in a rack of at least forty 20- x 150-mm test tubes moved by hand until the separation is complete. It is sometimes advantageous to collect small fractions early and large ones toward the end. Eluted components can be detected by spotting a sample from each fraction on a TLC plate. Collected fractions can then be appropriately combined and the solvent evaporated to isolate each component. After the separation has been completed, the silica gel in the column is washed with about 5 in. of ethyl acetate or acetone so that the column can be reused.

11.3 Flash Chromatography (Microscale)

Flash chromatography on a large scale requires specialized glassware and a source of air or nitrogen. On a very small scale, or *microscale*, a disposable Pasteur pipet can be used to hold the adsorbent in a flash chromatography procedure (Figure 11.3). Air forced through the column by pressure from a pipet bulb is sufficient to force the eluting solvents through the adsorbent. The following steps outline the purification of an organic compound by microscale flash chromatography.

(1) Select a Solvent System

Choose an elution solvent using the same procedure as for macroscale flash chromatography (step 1, p. 126).

(2) Prepare the Column

Plug a Pasteur pipet with a small amount of cotton. Take care that you do not either use too much cotton or pack it too tightly. You just need enough to prevent the adsorbent from leaking out.

Add dry adsorbent, usually silica gel 230–400 mesh, to a depth of 5–6 cm. A tiny beaker works well to pour the adsorbent into the column. Tap the pipet to pack the adsorbent, then apply pressure with a pipet bulb to pack it more. Recheck the depth and add more adsorbent if necessary so that the depth is 5–6 cm. This leaves a space of 4–5 cm on top of the adsorbent for the addition of solvent.

Fill the column with a nonpolar solvent, such as hexanes or petroleum ether, then force the solvent through the column until the liquid level is just flush with the top of the adsorbent. Repeat one time. This forces all air pockets from the column.

How to use a pipet bulb to force solvent through the column. Place the pipet bulb on top of the column, squeeze the bulb, and then remove the bulb while it is still squeezed. You must be careful not to allow the pipet bulb to expand before you remove it from the column, or you will draw solvent and adsorbent into the bulb. You will probably want to practice this technique on a sample column before you do it on your real experiment.

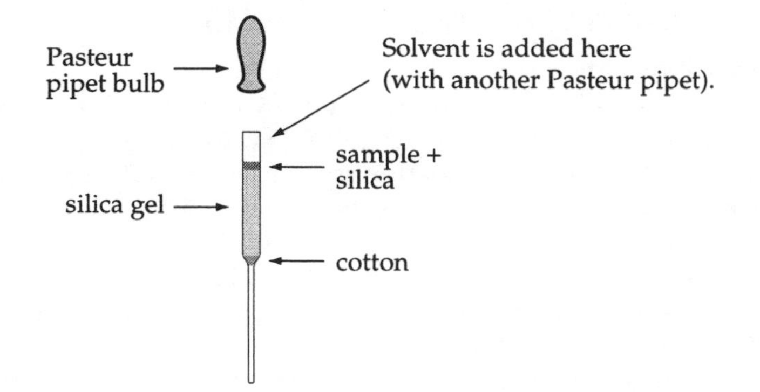

Figure 11.3 A Pasteur pipet as used in microscale flash chromatography. Pressure from the pipet bulb forces the solvent through the column.

(3) Load the Column

Weigh out 10–50 mg of sample and dissolve it in the minimum amount of eluting solvent. If it is not soluble in the eluting solvent, add 10–20% methylene chloride.

On occasion, the solvent used to dissolve and load the sample may be more polar than the eluting solvent. In this case, it is critical that you use only a few drops of solvent to load the sample. If you use too much, the loading solvent will interfere with the elution and hence the purification of the compound.

Another method for dealing with a sample that is not soluble in the eluting solvent is to "pre-load" the adsorbent. Mix the sample with about 150 mg of the adsorbent and enough of a polar solvent to dissolve the sample and to make a slurry. Then allow the solvent to evaporate completely, until the adsorbent is free-flowing. Do not heat the adsorbent mixture on a steam bath or it will pop and splatter! Transfer this adsorbent–sample mixture to the prepared column.

(4) Develop the Column

Place a vial under the column to collect the eluent, and have ready several more vials with which to collect subsequent fractions. (It's a good idea to have these vials pre-labeled with sequential numbers.) Fill the column with the elution solvent and then apply pressure with a pipet bulb until the level of the solvent is flush with the top of the adsorbent. Change collection vials and then repeat the process about 5-10 times.

If your compound is colored, you may not need to change vials with each application of eluting solvent. Instead, collect the eluent in a larger container, and change collection flasks as the colored compound elutes from the column.

Analyze each fraction by TLC unless otherwise specified. Combine fractions that contain pure samples of the compound and then evaporate the solvent.

11.4 High-Performance Liquid Chromatography (HPLC)

A relatively new chromatographic analytical procedure is high-performance (or high-pressure) liquid chromatography (HPLC). In HPLC, the solvent is pumped through the column at high pressure (6000 psi; 400 atm). The sample is "injected" into the solvent stream and passes through the column. The components in the sample are separated from one another as they pass through the column. The principles behind their separation are the same as those of column chromatography. As the components in the sample are eluted from the column, they pass through a detector, which converts their passage to electrical potential. The resulting signal is sent to a recorder, which then gives the chromatographic tracing.

The operation of HPLC and the interpretation of the results are similar to those for a gas chromatography (GC) instrument (Technique 13). The sample is injected and a tracing is obtained. The number of components in the sample can be estimated by counting the number of peaks in the chromatogram. The amount of each component is proportional to the area under each peak. However, HPLC is a far more powerful tool than GC and can be used to analyze a mixture that cannot be analyzed with gas chromatography.

Additional Resources

A list of suggested readings pertinent to column chromatography, expanded coverage of HPLC, and additional problems are published on the Brooks/Cole Web site. Please visit www.brookscole.com.

Problems

11.1 A chemist wishes to carry out a gravity chromatographic separation using diethyl ether and ethanol as the eluting solvents.

(a) With which solvent should the chemist begin the elution?

(b) What would happen if the chemist started with the other solvent?

11.2 A highly polar compound is moving through a gravity chromatography column too slowly. What can the chromatographer do to increase its rate of movement?

11.3 List the following compounds in order of expected elution from a chromatography column packed with silica gel, using a petroleum ether–diethyl ether solvent system.

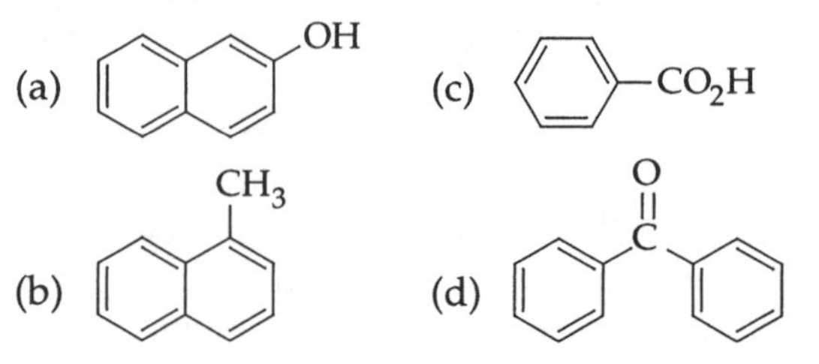

11.3 List the following compounds in order of expected elution from a chromatography column packed with silica gel, using a petroleum ether–diethyl ether solvent system.

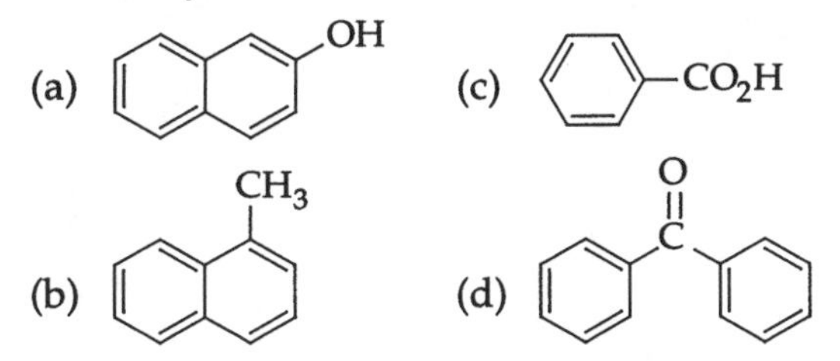

11.4 What will be the result of the following errors in a gravity chromatography process?

(a) Holes or air bubbles are present in the column packing.
(b) The column is loaded with too much sample.
(c) The column is eluted first with 5% diethyl ether in hexanes, then the eluting solvent is immediately switched to 50% diethyl ether in hexanes.
(d) The column is allowed to go dry before it is completely developed.

11.5 The R_f values determined on silica gel TLC plates for Compounds A, B, and C are tabulated in Table 11.5.

(a) In order to purify Compound A, which of the solvent systems should be used to develop the column in a flash chromatography procedure?
(b) Could 90 mg of a mixture of Compound A and Compound B be separated by macroscale flash column chromatography (10 mm column) using 20% ethyl acetate–80% hexanes as the eluting solvent?
(c) Could 90 mg of a mixture of Compound A and Compound C be separated by macroscale flash column chromatography (10 mm column) using 20% ethyl acetate–80% hexanes as the eluting solvent?

11.6 Of Compounds A, B, and C presented in Table 11.5, which is the most polar? The least polar?

11.7 Outline a procedure for separating Compound A from Compound C by flash chromatography using the data presented in Table 11.5. (Hint: First elute the less polar compound from the column and then elute the more polar compound.)

Table 11.5 The R_f values of three hypothetical compounds in four eluting solvents.

Solvent	*R_f of Cmpd A*	*R_f of Cmpd B*	*R_f of Cmpd C*
10% ethyl acetate–90% hexanes	0.12	0.07	0.01
20% ethyl acetate–80% hexanes	0.34	0.29	0.10
30% ethyl acetate–70% hexanes	0.53	0.42	0.22
40% ethyl acetate–60% hexanes	0.72	0.61	0.35

Thin-Layer Chromatography

Thin-layer chromatography (abbreviated TLC) is closely related to column chromatography. Instead of a column, the adsorbent is coated on one side of a strip or *plate* of glass, plastic, or aluminum. Instead of traveling *down* the adsorbent as in a column, the solvents and compounds travel *up* the plate by capillary action.

In the TLC analysis, a few microliters of a solution of the substance to be tested is placed ("spotted") in a single, small spot near one end of the plate using a microcapillary. The plate is "developed" by placing it in a jar with a small amount of solvent. Figure 12.1 shows a TLC plate in a developing jar. The solvent rises up the plate by capillary action, carrying the components of the sample with it. Different compounds in the sample are carried different distances up the plate because of variations in their adsorption on the adsorbent coating. If several components are present in a sample, a column of spots is seen on the developed plate, with the more polar compounds toward the bottom of the plate and the less polar compounds toward the top. (Table 11.1, p. 122, gives the expected order of elution of classes of organic compounds.)

As an analytical tool, TLC has a number of advantages: It is simple, quick, and inexpensive, and it requires only small amounts of sample. TLC is generally used as a qualitative analytical technique, such as checking the purity of a compound or determining the number of components in a mixture or column chromatographic fraction. We can use TLC to follow the course of a reaction by checking the disappearance of starting material and the appearance of product. In addition, TLC is useful for determining the best solvents for a column chromatographic separation. It can also be used for an initial check on the identity of an unknown sample (by spotting the plate with a known compound as well as with the sample). With calibration, TLC can be used as a quantitative technique. Preparative work can be carried out with special thick-layered TLC plates. TLC is fast, efficient, and simple to use. In all its forms, TLC is a very powerful tool.

12.1 The R_f Value

The distance that the spot of a particular compound moves up the plate relative to the distance moved by the solvent front is called the **retention factor**, or **R_f value**.

$$R_f = \frac{\text{distance traveled by the compound}}{\text{distance traveled by the solvent}}$$

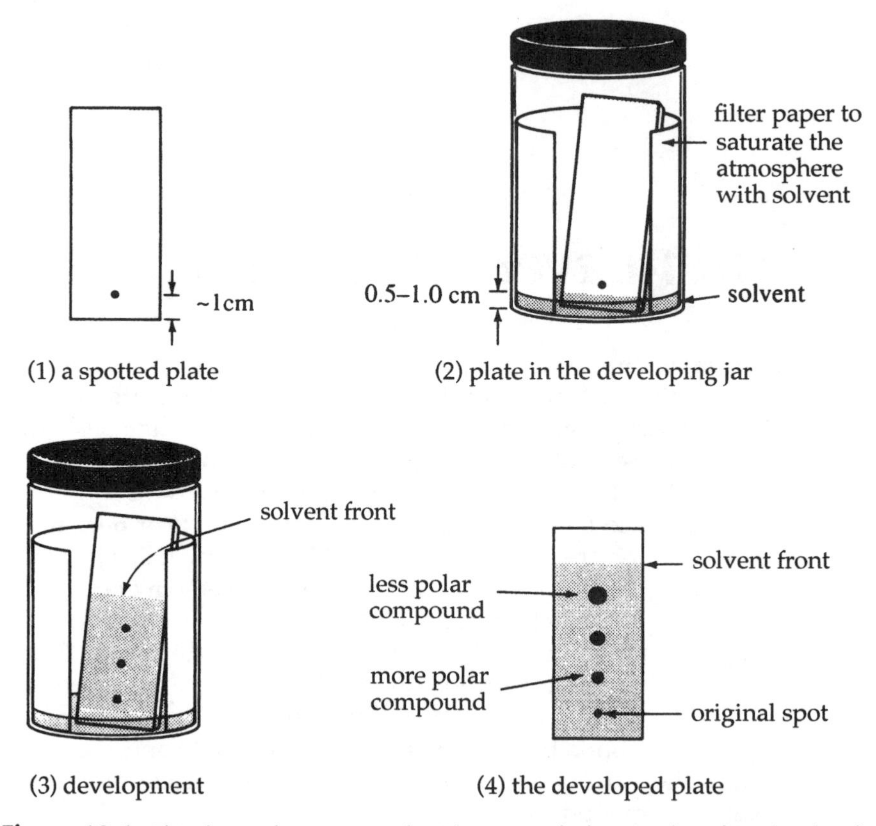

Figure 12.1 Thin-layer chromatography. The spotted plate is placed in the developing jar with a piece of filter paper, which acts as a wick to saturate the atmosphere with solvent. Different compounds move up the plate at different rates: The less polar compounds move the fastest and are found closer to the solvent front.

Figure 12.2 shows how these distances are measured. When the developed TLC plate is removed from the developing jar, the solvent front is marked immediately with a pencil before the solvent evaporates. Assuming that the compound spots are colored, the spots are outlined with a pencil in case the color fades. The distance that a compound has traveled is measured from the original spot to the center of the new spot. If the spot is elongated, the "center" is estimated (usually closer to the leading edge). The distance that the solvent has traveled is measured from the original spot to the solvent front.

The R_f value for a compound is a constant only if all variables are also held constant: temperature, solvent, adsorbent, thickness of adsorbent, amount of compound on the plate, and distance the solvent travels. Because it is difficult to duplicate all these factors exactly, an unknown sample is usually compared with a known compound on the same plate. Figure 12.3 shows how a mixture containing compound A compares with pure A on the same plate.

If two substances have the same R_f value, they are likely to be (but not necessarily) the same compound. A second TLC comparison using a different solvent for development may result in different R_f values, in which case the substances are not the same. If the second TLC analysis results in identical R_f values for the pair, the

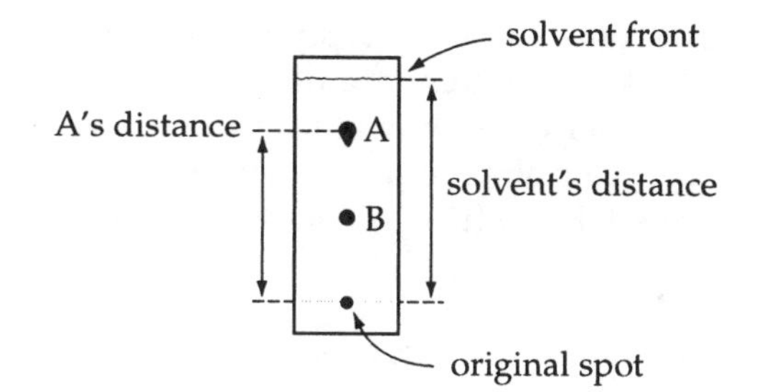

Figure 12.2 The R_f value for compound A is the ratio of the distance it has traveled to the distance the solvent has traveled. The spot for A is not circular here but shows "tailing"; therefore, the center of the spot is estimated. (Tailing is usually caused by too much sample in the original spot.)

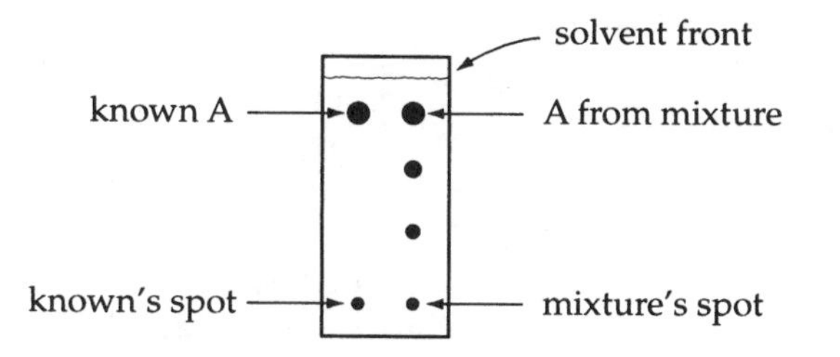

Figure 12.3 A known and an unknown sample should be analyzed on the same plate at the same time.

likelihood that the samples are identical increases. Even in this case, some other type of corroborating evidence is needed.

12.2 Equipment for TLC

(1) TLC Sheets and Plates

Commercial TLC sheets are coated with silica gel (SiO_2) or alumina (Al_2O_3). The properties of these adsorbents are discussed in Technique 11 (p. 121). Sometimes compounds are added to the adsorbent before it is spread on the plate. For instance, gypsum ($CaSO_4$, plaster of paris) is added to help it to bind to a glass plate. Silica gel with this binder is designated "silica gel G." Compounds that fluoresce under UV light (wavelength 254 nm) are commonly added. Silica gel products impregnated with such a fluor are designated "silica gel F_{254}."

(2) Pipets

Commercial 1-µL disposable micropipets, or *microcaps*, work well for spotting TLC plates.* The advantages of microcaps are that they are ready-made, each one is

* 1 microliter (µL) = 1×10^{-6} L, or 1×10^{-3} mL.

exactly the same size, and they can be rinsed and reused. If commercial microcaps are not available, draw out some soft glass tubing or melting-point capillary tubes in a flame. (CAUTION: Make sure there are no volatile solvents in the labroom before lighting a Bunsen burner.) The diameter of the drawn-out pipet should be about one-fourth of the diameter of a melting-point capillary. Homemade micropipets usually cannot be cleaned: Use a fresh pipet for each solution to be spotted, and then discard it after use.

(3) Developing Jars

Developing jars or chambers with the proper solvent system should be prepared well in advance of their use and kept in the fume hood. Any tall jar with a lid or screw top may be used for a developing jar, as can a beaker covered with a watch glass. The container should be narrow enough to hold the plate upright inside, without the danger of its falling over (see Figure 12.1). The lid of the container should be impervious to solvent fumes. The container should be small enough so that its atmosphere can be quickly equilibrated with solvent but large enough so that the sides of the plate do not touch the wick.

12.3 Steps in a TLC Analysis

(1) Preparing the Developing Jar

Line the inside of the jar halfway around with a piece of filter paper, which will act as a wick to saturate the atmosphere in the jar with solvent vapor. Before inserting the TLC plate, pour a small amount of the developing solvent into the jar to soak the filter paper and to cover the bottom of the jar to a depth of about 0.5–1.0 cm. The solvent level should cover the edge of the adsorbent on the plate and yet not reach the spots. Cap the jar and allow it to sit for at least 15 minutes to reach liquid–vapor equilibrium. Check that the solvent level is still about 0.5–1.0 cm, and add more solvent if necessary before inserting the plate. Only one plate can be developed in a jar at a time.

(2) Spotting the Plate

Prepare a TLC plate by drawing a light pencil line about 0.5–1.0 cm from one end to provide a guide for spotting the samples. Be careful that you do not press down too hard so that you do not disturb the adsorbent layer. The line needs to be far enough from the end of the plate so that it will be above the solvent level in the developing jar.

Dissolve about 1 mg of the solid or liquid sample in a few drops of a volatile solvent such as methanol or acetone. Dip the end of a micropipet into this solution, which will rise into the pipet by capillary action.

On the pencil line you have drawn on the plate, touch the end of the pipet *gently* and *briefly* to the adsorbent so that the solution runs out of the pipet and onto the adsorbent (Figure 12.4). Do not disturb the coating of adsorbent except where you are spotting. Make the spot as small as possible (1–3 mm in diameter) by allowing only a small amount of liquid to run out before lifting the pipet. As soon as the solvent evaporates, you may add more sample to the same spot. Depending on the concentration of the sample solution, one to three applications are usually sufficient. To determine the optimum number of applications, place three spots on one plate—the first spot containing one or two applications, the second spot containing three applications, and the third spot containing four or five applications.

It is important to spot the compounds high enough on the plate so that they will be *above the solvent level* in the developing jar. If the spots are below the solvent level, they will be dissolved off the plate by the solvent.

If more than one sample is being analyzed on the same plate, space the spots well apart and at the same distance from the bottom of the plate. Samples that are spotted too close together may spread out and run together as they are developed.

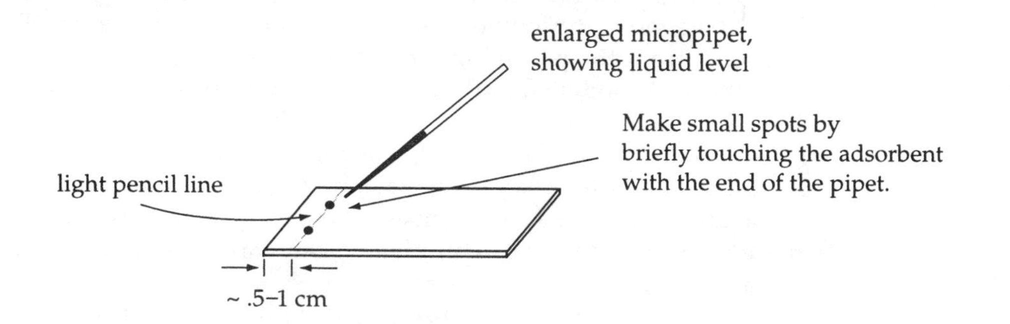
enlarged micropipet, showing liquid level
Make small spots by briefly touching the adsorbent with the end of the pipet.
light pencil line
~ .5–1 cm

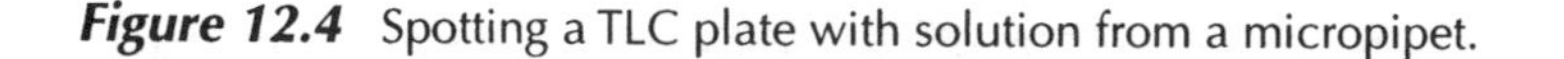

Figure 12.4 Spotting a TLC plate with solution from a micropipet.

(3) Developing the Plate

Prop the plate upright in the center of the jar (spots at the bottom) in such a way that the adsorbent side of the plate is visible through the side of the jar. Be sure the edges of the plate do not touch the wick. Cap the jar and do not move it during the development.

The solvent will rise up the adsorbent on the plate by capillary action. When the solvent front has risen almost to the top of the plate, open the jar, remove the plate, and quickly mark a line across the plate at the solvent front with a pencil. Check the plate for visible spots. Outline these carefully with the pencil, keeping your lines at the perimeters of the spots. A spot from a colorless organic compound will not be visible on the plate. Therefore, one or more visualization procedures may be followed.

(4) Visualizing the Spots

If the compounds are colored, simple observation of the plate after the plate has been developed is all that is required. However, as most organic compounds are not colored, one of several visualization methods is employed.

A common visualization method is that of observing the plates under ultraviolet light (also called a UV lamp or black light). Many aromatic compounds will show a dark blue spot under a UV lamp without any enhancement. Another technique consists of using a UV lamp to visualize spots on a TLC plate containing an inorganic fluorescent compound, such as zinc sulfide (ZnS), in its coating. When such a plate is placed under a black lamp, the entire plate glows. An organic compound capable of absorbing ultraviolet light or of quenching the fluorescence will show up as a dark spot.

A simple technique for visualizing spots is to place the dry, developed plate in a dry, covered jar with a few crystals of iodine (I_2). (CAUTION: Iodine is a strong irritant!) Most organic compounds show up as colored spots (yellow-brown to purple) when exposed to iodine vapor. The color arises either from the formation of a colored complex between the compound and iodine or from the dissolving of iodine in the compound. In either case, vapors of iodine are selectively adsorbed onto the TLC plate wherever there is a concentration of organic compound. If spots do not appear, warm the bottom of the jar gently (with your hand or briefly on a steam bath) to vaporize the iodine crystals. As soon as the spots are well defined, remove the plate from the jar and circle the spots with a pencil. This step is necessary because, as the iodine rapidly sublimes from the plate, the spots will become colorless again.

Another method for visualizing compounds is *charring*. The plate is heated to a high temperature in an oven, causing organic compounds to char or burn and show up as dark spots on the plate. Often, the plate is sprayed with sulfuric acid to hasten the charring. Note: This method can only be used if the backing of the plate is glass or aluminum—not plastic.

The use of *color-forming reactions* is another valuable technique for visualizing spots. A very specific procedure involves spraying the plate with a reagent that will react with a compound on the plate to form a colored product. For example, an acid–base indicator, like phenolphthalein, can be used to identify acidic or basic compounds. Other specialized sprays contain ferric chloride (to identify phenols), 2,4-dinitrophenylhydrazine, DNP (to identify aldehydes and ketones), and, with paper chromatography, ninhydrin (to identify amino acids).

12.4 A Related Technique: Paper Chromatography

In many respects, **paper chromatography** is similar to thin-layer chromatography. Instead of an adsorbent-coated plate, a strip of paper is used. Instead of a solid adsorbent, a thin film of water on the paper constitutes the adsorbent. Therefore, paper chromatography is a *liquid–liquid* partition technique, rather than a liquid–solid technique such as column chromatography and TLC.

For exacting analyses, commercial chromatographic paper strips equilibrated in a humid atmosphere should be used. In many analyses, however, filter paper can be used for paper chromatography because it is almost pure cellulose with few impurities. Under most atmospheric conditions, filter paper adsorbs moisture from the air. This adsorbed water makes up about 20% by weight of the filter paper and is usually sufficient for successful paper chromatography. Often, a damp solvent is used to develop the paper chromatogram to ensure the presence of sufficient water. A solvent that is fairly immiscible with water does not disturb the film of water adsorbed onto the cellulose.

Because very polar water molecules form the adsorbent layer, paper chromatography is most successful with very polar organic compounds. (Nonpolar compounds, which are not attracted to water, are usually carried with the solvent front.) Paper chromatography is commonly used for the identification of amino acids, which exist as highly polar dipolar ions, species containing full positive and negative ionic charges.

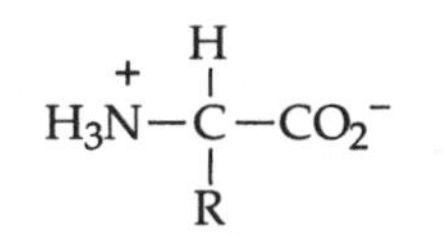

the dipolar ion of an amino acid

Additional Resources

A list of suggested readings for the technique of thin-layer chromatography is published on the Brooks/Cole Web site, as well as additional problems. Please visit www.brookscole.com.

Problems

12.1 Calculate the R_f values for the following compounds.
(a) Spot, 5.0 cm; solvent front, 20.0 cm
(b) Spot, 3.0 cm; solvent front, 12.0 cm
(c) Spot, 9.8 cm; solvent front, 12.0 cm

12.2 If two compounds have R_f values of 0.50 and 0.61, how far will they be separated from each other on a plate when the solvent front is developed to
(a) 5 cm?
(b) 15 cm?

12.3 A wick of filter paper is placed in a TLC developing jar, and the atmosphere in the jar is saturated with solvent before a plate is developed. What would happen if a plate were developed in a jar with an atmosphere not saturated with solvent vapor?

12.4 A student spots an unknown sample on a TLC plate. A single spot with an R_f of 0.55 showed up on the plate after developing the sample in hexanes–ethyl acetate 50:50. Does this indicate that the unknown material is a pure compound? What can be done to verify the purity of the sample?

12.5 As a separation and detection method, would TLC or paper chromatography yield better results in the analysis of each of the following pairs of compounds?

(a) $CH_3CH_2SO_3H$ and $CH_3CH_2SO_2NH_2$

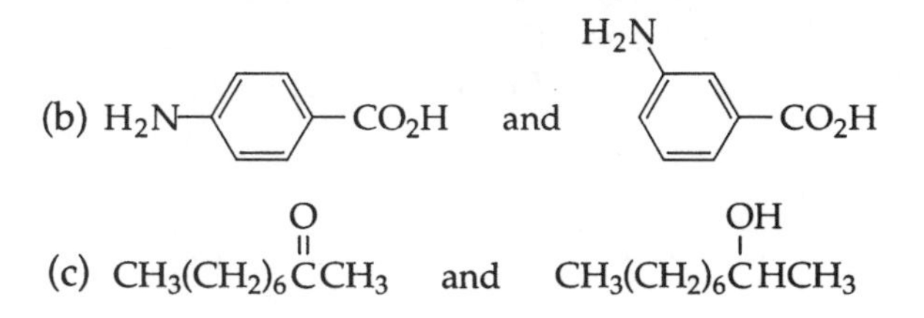

12.6 Consider a sample that is a mixture composed of biphenyl, benzoic acid, and benzyl alcohol. The sample is spotted on a TLC plate and developed in a hexanes–ethyl acetate solvent mixture. Predict the relative R_f values for the three compounds in the sample.

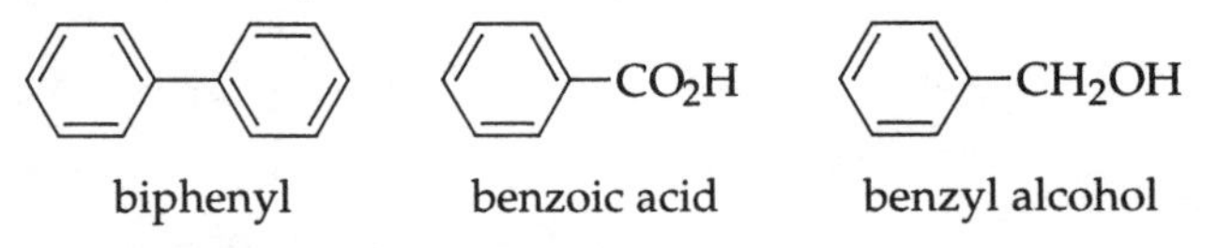

12.7 After a rather lengthy organic chemistry synthesis procedure, a student ran the product of the reaction on a TLC plate and obtained the result below. What might he/she have done wrong, if anything?

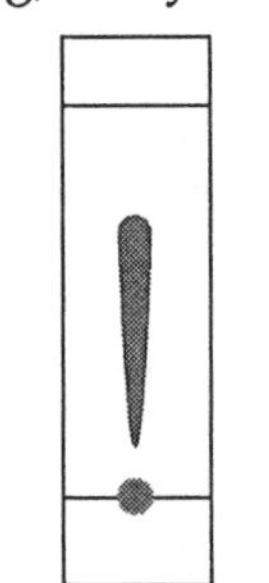

12.8 A student is following a reaction (the conversion of compound A to Compound B) by TLC. Aliquots of the reaction mixture are taken and analyzed at time = 0, 10, 20, and 30 min. The TLC plate, developed in 90% hexanes–10% ethyl acetate, is shown below.

(a) Which is the more polar compound, A or B?
(b) Is the reaction complete at 10 min? 20 min? 30 min?
(c) How can the identity of Compound B be verified by TLC?

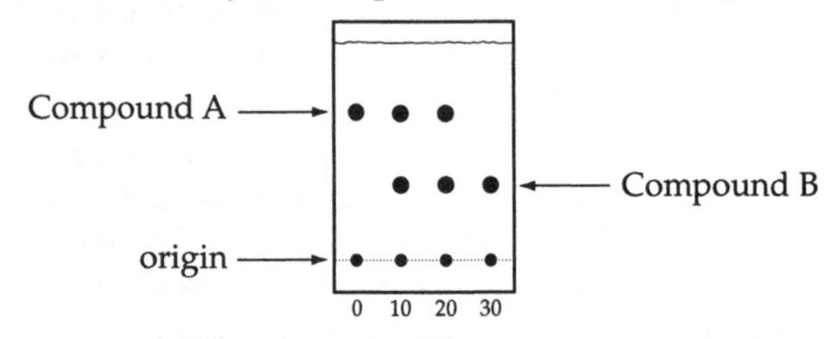

12.9 Consider the TLC plate illustrated in Problem 12.8. Would the R_f of Compound A be different if the solvent system used to develop the plate was 50% hexanes–50% ethyl acetate? If so, would it be lower or higher?

12.10 Again consider the TLC plate illustrated in Problem 12.8. Would the R_f of Compound B be different if the solvent system used to develop the plate was 100% hexanes? If so, would it be lower or higher?

Gas Chromatography

Gas chromatography (GC) is known by a variety of other names: gas–liquid chromatography (GLC), gas–liquid phase chromatography or gas–liquid partition chromatography (GLPC), and vapor phase chromatography. All these terms refer to the same technique, which is an instrumental method of analyzing the components of a mixture. In gas chromatography, the stationary phase is a high-boiling liquid and the mobile phase is an inert gas. Gas chromatography is generally used as an analytical tool rather than as a means of purification; however, like TLC, this technique can also be used to separate small quantities of compounds.

13.1 Gas Chromatography

The process of gas chromatography is carried out in a specially designed instrument, called a **gas chromatograph** (Figure 13.1). About 1–10 μL of liquid sample or solution is injected with a small hypodermic syringe into the gas chromatograph, through a rubber septum on the heated injection port of the instrument. The sample is vaporized and carried through a heated column by an inert carrier gas (usually helium or nitrogen). The adsorbent in the column is a high-boiling liquid suspended on a solid inert carrier. Because of differing interactions with the adsorbent and differing vapor pressures, the components of the sample move through the column at different rates. At the end of the column, each component passes through a detector, which is connected to a recorder. The recorder produces a tracing that shows *when* each component of a mixture passes the detector and also indicates the *approximate relative quantity* of each component.

GC is commonly used for checking the purity of volatile-liquid samples, such as distillation fractions; checking the identity of a substance by comparison of its GC with that of a known; and analyzing a mixture for the presence or absence of a known compound (such as alcohol in blood). In addition, the relative quantities of the components of a mixture can be determined with about 5% accuracy.

In most GC units, small quantities of sample are used and no attempt is made to collect the material that has been chromatographed. However, with *preparative gas chromatographs*, the separation and collection of samples are possible. Also, special techniques allow the effluents from a gas chromatograph to be further analyzed—for example, by a mass spectrometer.

As the vapor of a sample is carried through the column by the carrier gas, it continuously condenses and revaporizes. The amount of time the compound stays condensed depends on its volatility, and hence its boiling point. Once a compound

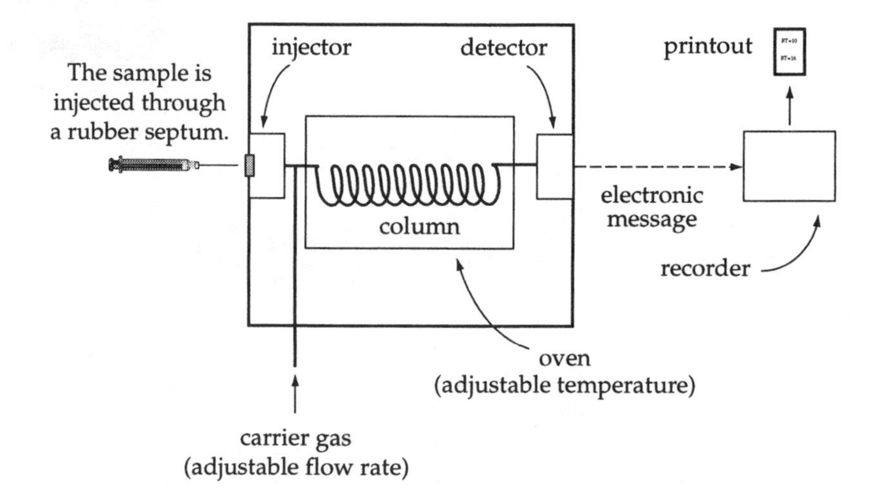

Figure 13.1 The basic components of the gas chromatograph. The vaporized sample is carried through the column by an inert carrier gas. The detector senses the passage of organic compounds, and the recorder graphs this information.

condenses, it can dissolve in the high-boiling liquid. The time it remains dissolved depends on its solubility in the liquid phase (according to the rule "like dissolves like"). Thus, a nonpolar compound with a high vapor pressure moves along a nonpolar column at a fairly rapid rate; a less volatile compound moves more slowly; and the compounds tend to concentrate in bands as they move through the column, as illustrated in Figure 13.2.

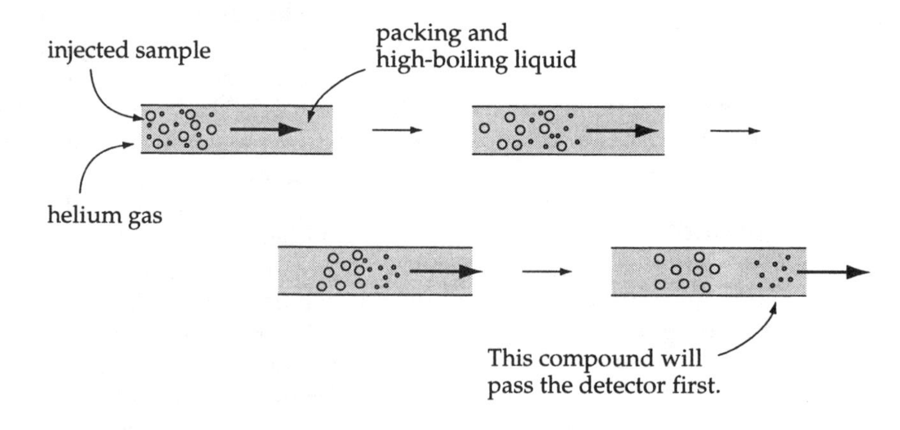

Figure 13.2 A mixture of compounds becomes separated into individual components as it passes through a GC column.

Columns in a GC can be changed, although in practice they usually are not changed very frequently. It is important to report which type of column is used in an experiment so that the experiment can be duplicated in another laboratory.

Efficient separation of compounds in GC is dependent on the compounds traveling through the column at different rates. The rate at which a compound travels through a particular GC system depends on all of the factors listed below.

- **Volatility of compound.** Low-boiling (volatile) components will travel faster through the column than will high-boiling components.
- **Polarity of compound.** Polar compounds will move more slowly, especially if the column is polar.
- **Temperature of the column.** Raising the column temperature speeds up all the compounds in a mixture.
- **Polarity of the column packing material.** Usually, all compounds will move more slowly on polar columns, but polar compounds will show a larger effect.
- **Flow rate of the gas through the column.** Speeding up the gas flow increases the speed with which all compounds move through the column.
- **Length of the column.** The longer the column, the longer it will take all compounds to elute. Longer columns are employed to obtain better separation.

If a GC trace indicates that the components are not separated in a particular run, several of the above factors can be varied to increase component separation. The first and easiest factor is to change the temperature of the column. Some instruments allow for the column temperature to be increased during a run, called a temperature *ramp*. The column itself can be changed, but this takes time. Flow rate is generally not changed from run to run.

13.2 The Gas Chromatograph

A. The Column

The column of a typical gas chromatograph is constructed of tubing, which is coiled so that a long column (3–10 ft) will fit into the small volume of the heated oven. The column is made of metal or glass and can range in diameter from capillary-size to 3/8-in. In general, the longer length and smaller diameter columns give better resolution of components but have a lower capacity. (The size column that can be used depends to some extent on the GC instrument itself.) The column contains an inert but loosely packed solid (such as crushed firebrick or diatomaceous earth) that is coated with the liquid phase. The inert packing provides spacing and allows the carrier gas to flow through the column without the need for excessive pressure. The liquid phase is the active component in the column. A number of interchangeable columns can be purchased for a gas chromatograph so that the appropriate packing liquid and column length can be chosen for a particular separation.

In the instrument, the column is located in an insulated oven with adjustable temperature controls. The upper useful limit of temperature is determined by the vapor pressure of the liquid phase in the column. The usual operating temperature is 100°–200°, but higher temperatures may be used with some types of columns. The

Table 13.1 Common stationary GC phases.

Name	*Composition*	*Polarity*	*Applications*
squalene and apiezon	hydrocarbon	nonpolar	saturated hydrocarbons
OV-1, OV-101	methylsilicone	nonpolar	amines, alcohols, ketones, alkaloids, hydrocarbons
SE-30	methylsilicone	low polarity	amines, alcohols, ketones, alkaloids, hydrocarbons
carbowax	polyethylene glycol	medium-high polarity	polar compounds: alcohols, ethers, amines, aldehydes, ketones
DEGS	diethylene glycol succinate	high polarity	polar compounds: esters, acids

operating temperature chosen depends on the boiling points of the compounds in the sample to be chromatographed.

A typical liquid phase used in a GC column is OV-101, which is useful for a wide variety of compounds. For polar compounds, a more polar liquid phase, such as a Carbowax, may be used. Columns containing less polar liquids are occasionally used for the GC analysis of nonpolar compounds. Typical liquid phases used in GC columns are listed in Table 13.1, and the structures of these compounds are given in Figure 13.3.

In columns that have a high-boiling, nonpolar hydrocarbon as the liquid phase (a nonpolar column), the separation of compounds in GC is similar to their separation in distillation, because both processes depend primarily on relative vapor pressures. The low-boiling components of a mixture pass through the GC column first, followed by successively higher-boiling components.

When the column contains a polar-liquid phase, the separation of the components in the mixture depends upon (1) their relative vapor pressures and (2) their relative interactions with the polar-liquid phase. The greater the vapor pressure, the faster the compound will pass through the column. However, the greater the interaction, the slower the compound will pass through the column. Because these two effects are unrelated (a compound can be low-boiling and polar, for example, or

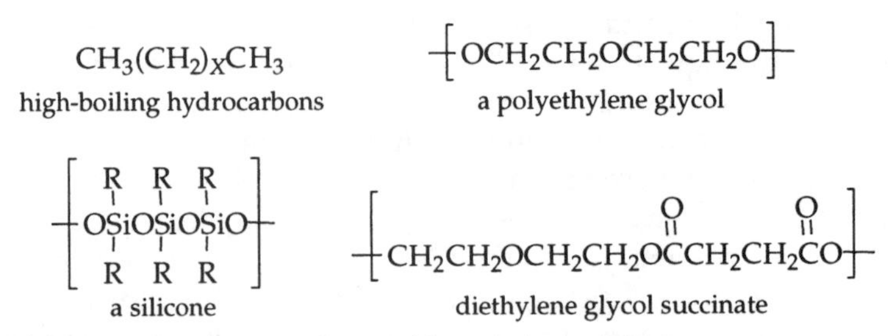

Figure 13.3 Structures of typical liquid phases in GC.

high-boiling and nonpolar), it is often very difficult to predict the order in which compounds will elute from the column.

B. The Detector

The detector at the end of the column signals when a compound is being eluted from the column. A number of different types of detectors have been invented, but only two types are in common use: **thermal conductivity (hot-wire) detectors** and **flame ionization detectors**.

In a thermal conductivity detector, an electric current is passed through a wire located directly in the flow of gaseous effluent from the column. The electrical resistance of a hot wire varies with the heat conductivity of a gas passing over it. Helium, with a high heat conductivity, absorbs heat from the wire and keeps it relatively cool. (Nitrogen can be used, but while it is much less expensive than helium, it has a lower thermal conductivity, making it less desirable for thermal conductivity detectors.) Organic compounds have lower heat conductivities. Therefore, when the vapors of an organic compound pass over the wire, the wire becomes hotter; consequently, its resistance changes. The hot wire is actually one arm (the sample arm) of a *Wheatstone bridge* circuit (see Figure 13.4). The other arm (the reference arm) is a similar hot wire with pure helium (no sample) passing over it. When no sample is passing through the sample arm, the bridge is in balance and no message is sent to the recorder. The passage of a gaseous organic compound from the column through the sample arm of the detector changes the resistance of the sample arm. Thus, the bridge becomes unbalanced, and an electric current passes to the recorder.

In a flame ionization detector (Figure 13.5), part of the gaseous effluent from the column is mixed with a separate gas stream of hydrogen and oxygen, and this gaseous mixture is burned. When organic compounds are present in the effluent from the column, they are oxidized and converted to ions by the high temperature of the flame. The ions then pass through a metal ring, creating an electrical potential difference between the ring and the barrel of the burner. The small potential difference is amplified and then sent to the recorder.

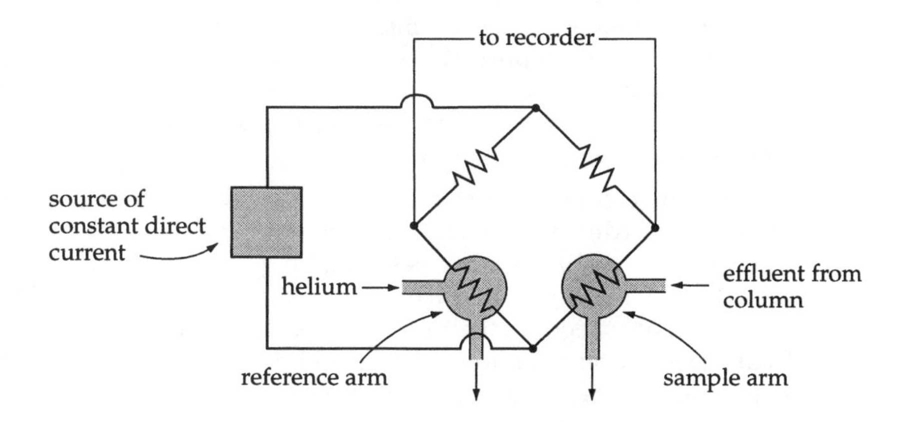

Figure 13.4 A Wheatstone bridge "hot-wire" detector.

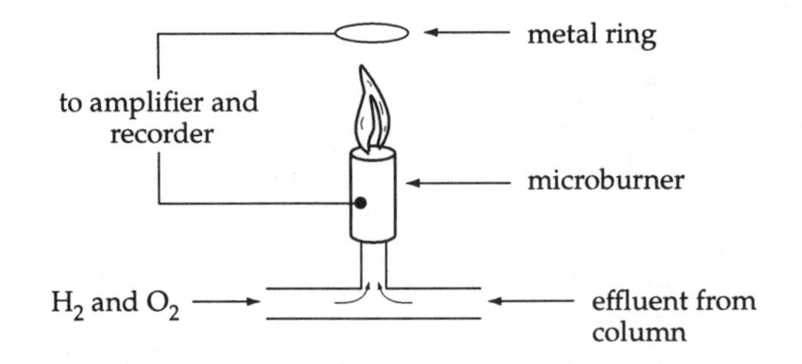

Figure 13.5 A flame ionization detector.

C. The Chromatogram

The **recorder** receives the electrical information from the detector and produces a graph—the **chromatogram**—of the components passing through the detector. Figure 13.6 is a representation of a typical gas chromatogram.

The time it takes for a particular compound to pass through the column is called the compound's **retention time (RT)**. The retention time is a function of the physical properties of the compound, the rate of gas flow, the temperature, the liquid phase, and the length and diameter of the column.

The retention time is measured from the injection point ("start") to the top of the compound peak and is usually reported in minutes. Under carefully controlled conditions, retention times can be duplicated. If accuracy is desired, an inert compound such as a hydrocarbon can be added to the sample and relative retention times measured from this internal standard's peak instead of from the injection point. With this technique, relative retention times can be reproduced with an accuracy of about 1%.

Undergraduate teaching laboratories are equipped with one of several styles of GC recorders: integrating recorders, chart recorders, or computer interfaces.

Integrating recorders. These instruments record the electronic input from the detector as peaks and they also integrate the area under each peak in the chromatogram. The printout reports the number of peaks, the retention time of each peak, and the area under each peak (relative to the areas of the other peaks). To operate the instrument, you press "start" on the keypad as you inject, and you press "stop" when you have seen the expected peaks.

Chart recorders. Older-style chart recorders feature a roll of paper that rolls under a recorder pen. When the detector sends an electronic message to the recorder pen, the pen moves, resulting in a "peak" on the chart paper. Students need to mark the chart paper when they inject their sample (the injection point) and note the chart speed. Retention times are determined by measuring the distance from the injection point to the peaks and dividing this measurement by the chart speed. Areas under each peak are determined by *triangulation* (see p. 150).

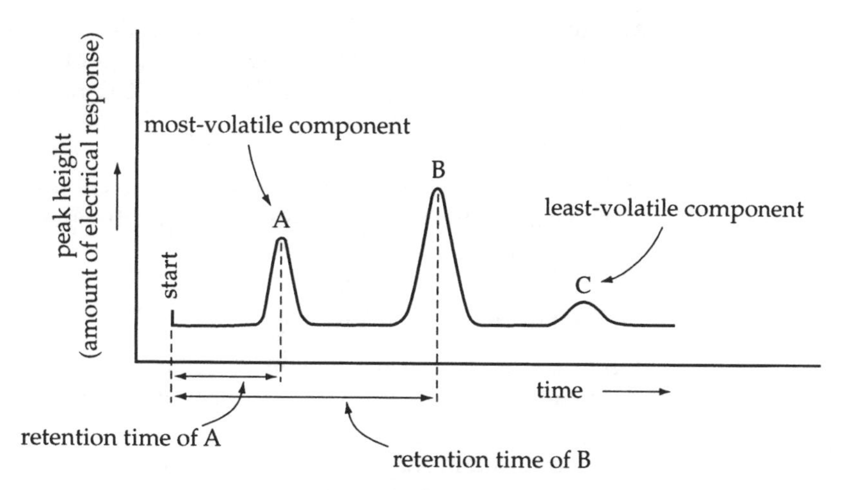

Figure 13.6 A typical gas chromatogram. The retention time of a compound is measured as the distance between the start point and the top of the compound's peak. The areas under the peaks are determined by the integrating recorder, the computer program, or manual triangulation.

Computer interface. Your laboratory instructor will provide instructions for use of the GC–computer interface. Retention times and areas under the peaks will be displayed on the screen and then printed out on a sheet of paper.

13.3 Steps in a GC Analysis

(1) Preparing the Instrument

A gas chromatograph cannot be turned on and used immediately. The injection port, the detector, and the oven must be allowed to come to thermal equilibrium. For this reason, one or more of the heat sources (depending on the instrument) and a slow flow of carrier gas should be started well in advance of instrument use. Your instructor will probably prepare the instrument before class.

Each time you make a run on the instrument, make sure that the gas flow is adjusted properly, the temperature is at the proper level, and the auxiliary equipment (detector or recorder) is switched on. Your instructor will demonstrate these manipulations.

(2) Injecting the Sample

Use a microhypodermic syringe to inject a sample through the silicone rubber septum into the vaporizing chamber. The amount to inject depends on the capacity and sensitivity of the instrument; a few microliters is usually sufficient. (CAUTION: Microsyringes are delicate and expensive. The metal plunger and needle are easily bent. Dirt in the glass barrel of the syringe will cause the plunger to stick; if the plunger is forced, the barrel will split. Your instructor will demonstrate the proper handling and cleaning procedures for microsyringes.)

Inject the sample quickly, so that it vaporizes all at once. (CAUTION: The injection port is very hot—do not touch it!) With a larger microsyringe (100 μL), hold the plunger before and after the injection. (The carrier gas in some instruments is under enough pressure that it can cause the plunger to fly out.) Mark the injection point according to instructions (each style of GC recorder requires a different method).

If a solution (instead of an undiluted sample mixture) is injected, a large solvent peak will be observed at the start of the chromatogram. Do not worry if the recorder pen runs to the top of the recorder paper; the pen will return to the baseline when all the solvent has passed the detector.

(3) Obtaining the Chromatogram

After making an injection, an operator has little to do but watch the tracing of the pen. Do not make another injection until the previous one has passed through the column.

After the chromatogram has been recorded, stop the recorder according to the instructions in your laboratory. Record the column type (silicone, Carbowax, and so on), the temperature of the column, the flow rate of the carrier gas, and any comments you feel necessary.

(4) Calibrating Gas Chromatograms

There are two ways to calibrate a gas chromatogram to identify components. The first is to run a chromatogram of a pure sample of known composition immediately before or after the chromatogram of the unknown sample. The conditions (temperature, gas flow, and so on) should not be changed during the calibration run and actual run. The second calibration method assumes that a sample has already been subjected to GC analysis, so that the appearance of the gas chromatogram is known. A drop of the known is mixed with a few drops of the sample, and the chromatogram of this mixture is then compared with the original chromatogram. If the addition of the known has increased the size of one of the existing peaks, the compound giving rise to the peak may be identical with the known. If the addition of the known compound to the mixture results in a new peak, then that known compound is not present in the sample. While such identification techniques are useful in the laboratory, they are not considered an absolute proof of structure. Corroborating evidence is needed.

Safety Notes

- Cylinders of compressed gas must always be chained to the wall or laboratory bench so that they cannot be knocked over.
- Do not handle the valves on a cylinder of compressed gas until their use has been demonstrated.
- The GC injector port is hot—do not touch it!

13.4 Problems Encountered in GC

Most of the problems encountered in GC analysis are easily correctable. *Poor resolution*, in which peaks overlap each other, may be the result of too large a sample, too high a gas flow, or too high a temperature. In some cases, a different type of column (for example, Carbowax instead of silicone oil) or a longer column with a narrower diameter may be necessary for good resolution.

A peak that runs off the top of the recorder paper is the result of too large a sample or too high a sensitivity setting. Most instruments have a sensitivity control, called an *attenuator*, that can be adjusted if necessary. Increasing the attenuator setting decreases the peak height. If the peak is too high, either reinject with a smaller volume or increase the attenuator and reinject using the same volume.

A peak that is very broad or unsymmetrical can result from the injection of too much material. A compound with an unusually long retention time may also be broad and sometimes very flat. In this case, the gas flow should be rechecked. Increasing the temperature or switching to another type of column usually solves this problem.

Some problems in GC work can be traced to a worn-out injection septum. The carrier gas and the sample can both leak out through holes or cracks in the septum. Leakage leads to small sample peaks (or none at all) and nonreproducible retention times.

If an injected sample gives rise to no recorder signal but the instrument is otherwise operating normally, the sample may have been retained in the column. A retained sample can result in contamination of future analyses, a change in the character of the column, and even a plugged column. For these reasons, avoid injecting tars or polymers into a GC column. (In fact, your instructor may ask you to analyze only *distilled* samples by GC.) Notify your instructor if your sample does not pass through the column.

13.5 Uses of Gas Chromatograms

A. Qualitative GC Analysis

As we have mentioned, GC is commonly used for checking the purity of a volatile sample. Determining the actual identity of a sample component from a GC analysis cannot be done directly. In a research laboratory, it may be necessary to isolate a compound, such as by preparative GC, and to determine its structure by physical constants, spectral data, and chemical analysis. In routine laboratory analyses, when we already have a good idea of the components in a mixture, such a lengthy identification procedure is usually not necessary.

In some cases, the structure of a compound giving rise to a peak in a GC chromatogram can be determined indirectly. If the compound is known to belong to a particular homologous series, it is possible to identify the compound by comparing the log of its retention time with those of three (or more) other members of the same homologous series. For example, suppose that the sample compound is a methyl ester of a carboxylic acid with a continuous, saturated chain of unknown length. We would select three known methyl esters of continuous-chain carboxylic acids and plot the logs of their retention times versus the number of carbons in their chains. The result is a straight line. The number of carbons in the chain of the unknown ester can then be determined from the graph and the log of its retention time.

Besides showing the minimum number of components in a sample and their retention times, a gas chromatogram obtained from a nonpolar column can show the approximate relative boiling points. If the instrument is calibrated with similar compounds of known boiling points, we can closely estimate the actual boiling points of the components in the sample.

B. Quantitative Analysis

The area under a peak in a gas chromatogram is directly proportional to (1) the amount of compound in the sample and (2) the detector response. Detector response varies with different types of compounds. If the variations in detector response are ignored, we can calculate the compound's percent easily, but our accuracy is not great (around 10%). Accuracy can be improved by calibrating the instrument by injecting known amounts of the compounds and calculating a correction factor for detector response.

Determining peak areas. Regardless of whether or not the chromatogram is to be corrected for detector response, peak areas—not peak heights—must be used for quantitative calculations. Peak heights depend on retention time; peak areas do not. Many gas-chromatographic recorders are equipped with mechanical or electronic **integrators** that automatically provide peak areas. If your recorder has one of these devices, your instructor will show you how it is used.

If a gas chromatograph does not have an integrator, the area can be approximated with an accuracy of about 3% by considering the peak to be a *triangle* (area = 1/2 base x height). Instead of measuring the base of a GC peak and dividing by 2, use the width of the peak at 1/2 peak height, because it yields more reproducible values. Determining areas by triangulation works well only if the peaks are well resolved and symmetrical.

Another technique for determining the relative areas under peaks, including unsymmetrical peaks, is to carefully cut out the peaks (or tracings of the peaks on plain, uniform bond paper) and weigh them. The weights of the peaks are proportional to the areas.

Quantitative analysis without correcting for detector response. If a rough approximation of the percents of each component in a sample is satisfactory, then the area of each component is expressed as a percent of the total area of all components in the chromatogram. Because of its simplicity, this is the method most commonly used in quantitative GC analysis.

Correcting areas with an internal standard. The weight ratios or weight percents of components in a mixture can be determined indirectly from a gas chromatogram by considering the *areas under the peaks* and *correction factors* that compensate for different detector responses. Although there are several ways to convert area ratios to corrected weight ratios, we will present only one: the **internal-standard technique**. A detector produces a signal that is directly proportional to the quantity of compound injected. Because different compounds exhibit different proportionality constants, it is necessary to obtain correction factors for quantitative determinations of mixture composition. The internal-standard technique utilizes weight and area ratios for these corrections.

In the internal-standard technique, a reference compound, or standard, is added to both a calibrating mixture and the mixture to be analyzed. Any compound can be used as a standard. However, for best results, the standard should be structurally similar to the components being analyzed and should be present in the mixture in approximately the same concentration. Also, the instrument should be adjusted so that the peak of the standard is well separated from the other peaks in the chromatogram.

The internal-standard technique involves three general steps. First, the calibrating mixture of *known weights* of the components to be analyzed and the standard is prepared; the chromatogram of this known mixture is then obtained. Next, the weights and peak areas for this known mixture are used to calculate a correction factor for each component, which can be applied to a chromatogram of a mixture of unknown percent composition. Finally, the chromatogram of the mixture of unknown percent composition (also containing the reference compound) is obtained, and the percent composition of this mixture is calculated.

Additional Resources

A list of suggested readings for the technique of gas chromatography, examples of calculations for correcting areas with an internal standard, and additional problems are published on the Brooks/Cole Web site. Please visit www.brookscole.com.

Problems

13.1 Why are the following procedures invalid?

(a) To minimize the time required for an analysis, the temperature of a GC instrument is increased above the boiling points of the components in the mixture.

(b) To increase the retention times and thus maximize the separation between peaks, the gas flow is adjusted to a very low value.

13.2 Alcohols have very long retention times on Carbowax columns. Why? (Hint: Consider the polarity of alcohols.)

13.3 Suppose an organic compound has a higher heat conductivity than the carrier gas. How would its GC signal appear on a chromatogram run on an instrument with

(a) a thermal conductivity detector?

(b) a flame ionization detector?

13.4 The RT for cyclohexane on a GC instrument is 0.75 min. The temperature of the GC is 70°C and the gas flow rate is 60 mL/min.

(a) Would the RT be higher or lower if the column temperature is raised to 100°C?

(b) Would the RT be higher or lower if the flow rate is changed to 30 mL/min?

13.5 Between sample injections, some students clean a microsyringe with acetone rather than with the mixture to be analyzed. Why is this a poor technique?

13.6 Helium is the carrier gas of choice for a gas chromatograph containing a thermal conductivity detector, and nitrogen gas is preferred for a gas chromatograph having a flame ionization detector. Suggest a reason for this.

13.7 Suppose that, on a GC column, methyl hexanoate eluted in 1.87 minutes, methyl heptanoate eluted in 2.35 minutes, and methyl octanoate eluted in 3.0

minutes. What is the structure of an unknown methyl ester, known to belong to the same homologous series, that elutes at 4.7 minutes?

13.8 A student at another university wants to duplicate a GC separation that you did in lab. What information about the instrument and column conditions should you send him?

13.9 Consider the following compounds:

Compound	*Boiling point*
methyl cyclohexane	101°C
pentane	36°C
octane	126°C
2,3-dimethyloctane	165°C
heptane	98°C

If a mixture of these compounds is run on a GC using an OV-101 column, which compound will have the lowest RT? the highest?

13.10 Consider the following compounds:

Compound	*Boiling point*
propionic acid	141°C
2-hexanol	140°C
isoamyl acetate	142°C
3,4-dimethylheptane	140°C

If a mixture of these compounds is run on a GC using a DEGS column, which compound will have the lowest RT? the highest?

Carrying Out Typical Reactions

In the laboratory, an organic chemist often conducts an experiment or a synthesis by the following series of steps:

(1) Carry out the reaction.

(2) Isolate the product, perhaps by extraction, crystallization, distillation, or a combination of these techniques.

(3) Purify the product, perhaps by crystallization, distillation, sublimation, or a chromatographic technique.

(4) Characterize the product and estimate its purity by one or more of the following techniques: melting-point, boiling-point, refractive index, gas chromatography, thin-layer chromatography, spectroscopy, or elemental analysis.

Steps (2) and (3), isolating and purifying the product, are often referred to as the **workup** of the reaction mixture. How a reaction mixture is worked up depends on the physical and chemical properties of the product and by-products. Many of the techniques used to isolate, purify, and characterize an organic compound have already been presented. In this technique, we will discuss only step (1) in the preceding list: how typical organic reactions are carried out.

An organic reaction is carried out in a *reaction vessel*, generally a flask (Erlenmeyer, round-bottom, or three-neck) or, for microscale reactions, a conical vial or a heavy-walled test tube. Choosing and equipping the vessel depend on the reaction itself, which may require stirring, heating, or cooling, and the addition of reagents during the reaction's course. Therefore, let us consider these operations in light of the equipment required.

14.1 Stirring

Homogeneous reaction mixtures need not be stirred, even when heated, because the reactants are in intimate contact. However, to prevent bumping, boiling chips must be added to a homogeneous unstirred mixture before it is heated to boiling. Heterogeneous reaction mixtures, or mixtures that may become heterogeneous, require stirring to mix the reactants. Stirring is particularly important when a heterogeneous mixture is heated, because stirring prevents bumping. (Boiling chips do not prevent bumping of heterogeneous mixtures containing solid materials.)

The simplest stirring technique is *swirling*. Swirling should be used only to mix solutions at, or near, room temperature. A flask containing a boiling liquid should not be swirled, because hot vapors from the flask can burn the hand. When the reaction vessel is an Erlenmeyer flask or a beaker, another simple stirring technique is hand stirring with a *glass rod* (not a thermometer).

A **magnetic stirrer**, used with a beaker or flask, accomplishes simple mixing mechanically. A stir bar, generally a ferrous metal encased in an inert Teflon coating, is placed in the reaction vessel. Stir bars come in many sizes and shapes, including very small sizes (called *spin vanes*) appropriate for the microscale reactions. The vessel is then placed on the motor casing. When the motor is turned on, it rotates a magnet, which causes the stir bar to turn (see Figure 14.1). The principal disadvantage of magnetic stirring is that sufficient torque cannot be generated for stirring a thick mixture.

Some magnetic stirrers are equipped with hot plates so that mixtures can be warmed and stirred simultaneously. When heating a flask in a water bath while using one of these devices, you must use a nonferrous pan (aluminum or glass). Similarly, a cooling bath used with a magnetic stirrer must be nonferrous.

The stirrer of choice is a **mechanical stirrer** powered by a sparkproof motor. This type of stirrer, which is commonly found in research laboratories but rarely in student laboratories, is sufficiently powerful to stir even viscous mixtures. Figure 14.5 (p. 160) depicts a mechanical stirrer. The stir blade at the bottom of the heavy glass rod is made from glass or Teflon and can be changed to fit the size of the flask.

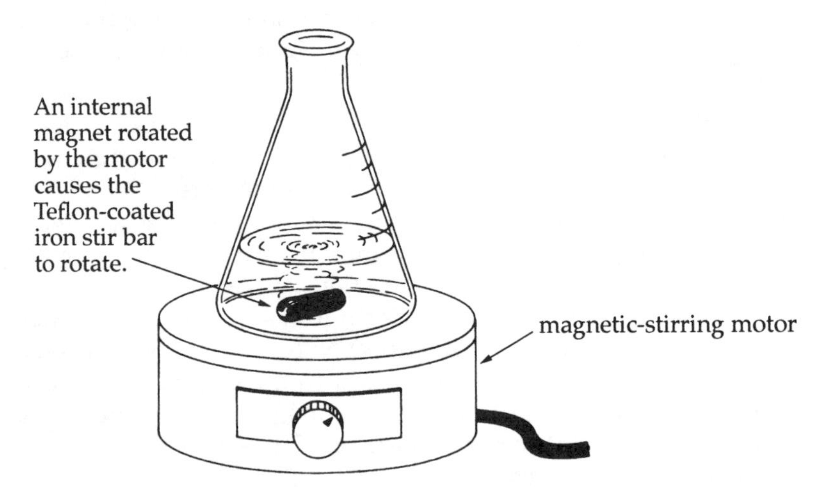
An internal magnet rotated by the motor causes the Teflon-coated iron stir bar to rotate.
magnetic-stirring motor

Figure 14.1 A magnetic stirrer can be used to stir mixtures that are not viscous.

14.2 Heating Reaction Mixtures

Most organic reactions require that the reactants be heated. As in distillation, the heat source chosen depends on the temperature that must be attained and the flammability of evolved vapors. Thus, a steam bath, a hot plate, a heating mantle, a heat lamp, or (rarely) a burner might be chosen. For a reaction that must be kept very dry, a heating mantle is preferable to a steam bath. Regardless of the heat source, the heat should be applied evenly to avoid localized overheating. (The use of heating devices is discussed in the introductory chapter; see p. 15.)

Boiling chips. Boiling chips, or boiling stones, should be used whenever a solvent is brought to a boil, unless the liquid can be constantly stirred or swirled. Boiling chips are small, porous stones of calcium carbonate or silicon carbide that contain trapped air. When the chips are heated in a solvent, they release tiny air bubbles, which ensure even boiling. When all the air has been released, the stone's pores provide cavities where bubbles of solvent vapor can form. Without boiling chips, part of the solvent may become superheated and boil in spurts, a process called *bumping*. Bumping is likely to cause some of the solution to splash out of the flask. Only two or three boiling chips are necessary to prevent bumping.

Boiling chips should never be added to a hot solution! If a solution is at or near its boiling point when boiling chips are added, the solution will almost certainly boil out of the flask.

Boiling chips will maintain their function throughout a long boiling period. Once they are used and cooled, however, the pores fill with liquid and lose their ability to release bubbles. Therefore, use fresh boiling chips each time you heat a solution.

Boiling sticks may be used instead of boiling chips. Propped upright in a flask, these small wooden sticks provide a porous surface on which the solvent bubbles can form. As with boiling chips, a fresh stick must be used each time the solution is heated.

Heating under reflux. "Reflux" means "a flowing back." In the chemistry laboratory, the term *reflux* refers to the return of condensed vapors to their original vessel.

Figure 14.2 shows a macroscale reaction assembly for heating a mixture under reflux. The water-cooled **reflux condenser** is an ordinary condenser arranged in an upright position so that vapors from the boiling liquid are condensed and returned to the flask. A reflux condenser can be used during spontaneous exothermic reactions or when a liquid is being boiled by a heat source. The purpose of reflux is twofold: A reaction can be maintained at the temperature of the boiling solvent, and solvent is not lost to the atmosphere.

In microscale reactions, a water-cooled reflux condenser is usually not necessary. Any sort of vertical extended glass tube above the reaction flask is sufficient to cool and condense the small amount of vapors used in microscale reactions. Some microscale glassware employs tall *reaction tubes*, where the reaction takes place, vaporizes, and condenses in one long, thick-walled test tube. Other styles of microscale glassware use *air condensers* that connect to the reaction vessel.

When heating a compound under reflux, keep in mind that heating too strongly will drive solvent vapors out the top of the condenser. A boiling reaction mixture is no warmer, and the reaction will not proceed any faster, when unnecessary heat is applied. Therefore, control the rate of heating so that the reflux level is in the *lower* portion of the condenser. If a low-boiling solvent is being heated, a second reflux condenser, placed on top of the first one, helps control the escape of vapors.

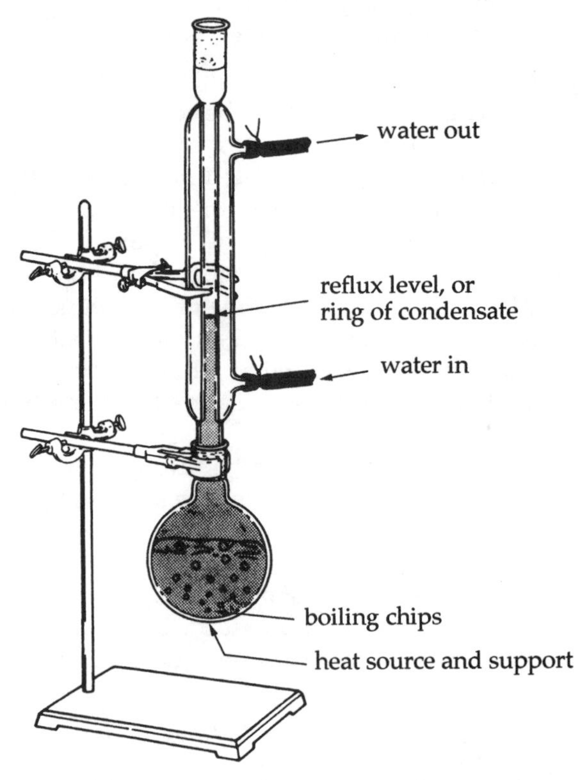

Figure 14.2 A round-bottom flask fitted with a reflux condenser.

14.3 Controlling Exothermic Reactions

Many reactions in your laboratory course are *exothermic* reactions—reactions that produce heat. If a reaction is highly exothermic, its temperature must be controlled. Otherwise, the reaction can "run away," that is, proceed too rapidly and boil too vigorously, even to the point of spewing out the top of the condenser! A few typical exothermic reactions are oxidation reactions, reactions using sodium metal, and Grignard reactions.

The rate of an exothermic reaction is controlled in three ways, which are usually used in conjunction with one another.

(1) *Regulation of the temperature,* usually by using a pan of ice water to chill the reaction flask. (You should always have an ice bath at hand when running an exothermic reaction.) Table 14.1 lists some other cooling baths and their minimum temperatures.

(2) *Regulation of the rate of addition* of one of the reactants. (Techniques for controlling the rate of addition are discussed in Section 14.4.)

(3) *Continuous stirring or swirling* during the addition so that "pockets" containing an excess of a reactant do not develop. (Thorough mixing is especially important if the reaction mixture is thick.)

Table 14.1 Some common cooling baths.

Cooling bath	*Minimum temperature*
ice–water	0°C
ice–salt water*	–21°C
ice–calcium chloride†	–54°C
dry ice–ethanol	–72°C
dry ice–acetone	–77°C

* 100 g ice–33 g NaCl
† 70 g ice–100 g $CaCl_2 \bullet 6H_2O$

14.4 Adding Reagents to a Reaction Mixture

In many reactions, the reactants are simply mixed and the mixture is heated under reflux until reaction is complete. In many other reactions, one reactant must be added slowly to the reaction mixture without allowing noxious fumes and vapors to escape.

A. Addition of Liquids

Small amounts of a liquid can be added to a reaction mixture by dropping them down a reflux condenser. The preferred technique in macroscale reactions, however, is to add the liquid from a dropping funnel. Figure 14.3 shows a simple reaction assembly equipped with a dropping funnel. (This figure also shows a *drying tube,* through which air can pass; see Section 14.5.) Note that the reaction flask is topped with a Claisen head and that the dropping funnel is positioned directly above the reaction flask. This positioning allows liquid to be added directly to the reaction mixture without its running down the sides of the glass. The other arm of the Claisen head is connected to a reflux condenser so that vapors can be condensed and returned to the reaction flask. A dropping funnel cannot be placed directly atop the reflux condenser as shown in Figure 14.2, because this would constitute a closed system.

Dropping funnels are generally not used in microscale reactions. Liquids are added to the reaction mixture with a syringe (or a pipet) through the reflux condenser or through one arm of a small Claisen adapter.

B. Addition of Solids

Adding solids to a reaction mixture is generally more difficult than adding liquids. If the solid is granular (not powdery) and not likely to stick to the wet inner sides of a reflux condenser, it may be feasible to drop the solid directly through the reflux condenser. However, a solid must usually be added directly to the reaction mixture without coming into contact with a glass surface. One way to add solids is to cool the reaction mixture, remove the reflux condenser, and add the solid to the flask itself. Alternatively, set up the reaction assembly shown in Figure 14.3, but with a ground-glass stopper in place of the dropping funnel. Then the solid can be added

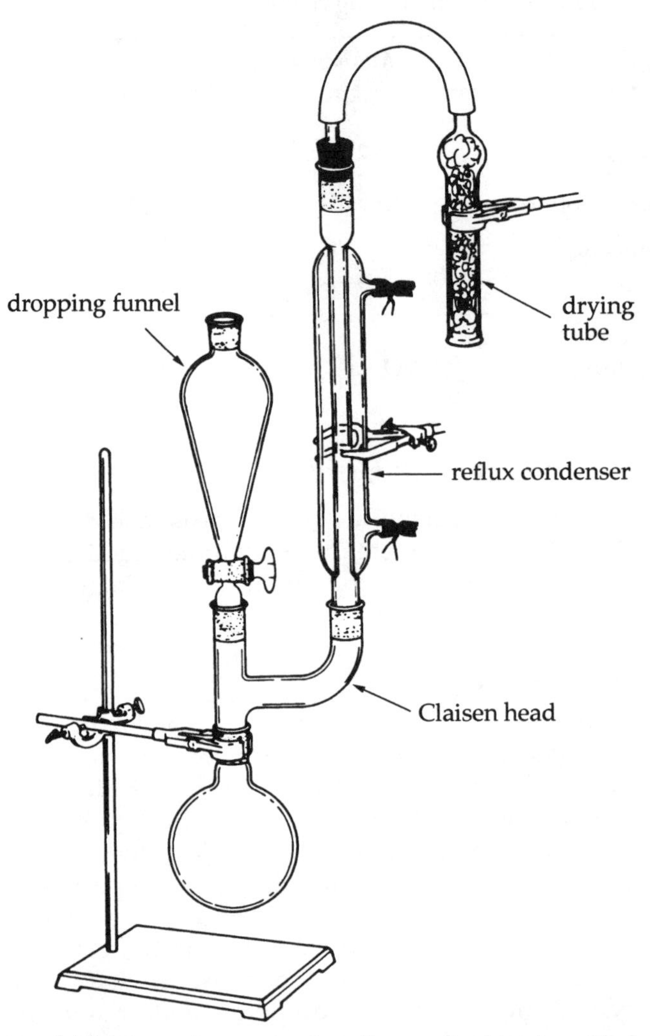

Figure 14.3 A reaction set-up that allows a liquid to be added to a mixture under reflux.

through this joint. When adding a solid through a ground-glass joint, always use a powder funnel. The wide stem will allow the solid to pass without clogging and will protect the ground-glass joint from contamination.

14.5 Excluding Moisture from a Reaction Mixture

Atmospheric moisture tends to condense inside a cold condenser. Some organic reactions require that moisture be excluded from their atmosphere. In a research laboratory, a chemist might exclude both oxygen and moisture by running a reaction under an atmosphere of a dried inert gas such as nitrogen or argon. This is seldom feasible in a student laboratory. Instead, a **drying tube** containing coarse pieces of a desiccant is used in both macroscale and microscale procedures. Anhydrous calcium chloride is the most common desiccant used in drying tubes.

The purpose of the drying tube is to prevent atmospheric moisture from entering the reaction vessel via the condenser and yet to allow the reaction vessel to be open to the atmosphere so that gas pressure does not build up. There are two types of drying tubes: curved (better) and straight (less expensive). A straight drying tube must not be connected directly to the top of the condenser, because the desiccant can liquefy and drain into the condenser. Connect the straight tube to the condenser with a thermometer adapter or with a short length of heavy-walled rubber tubing (Figure 14.4). The desiccant is held in place with loose plugs of glass wool. A one-hole rubber stopper may be used as a secondary plug at the wide end of the drying tube.

The length of time the desiccant in a drying tube remains sufficiently dry to be effective depends on the relative humidity and on the reaction vapors to which it is subjected. You cannot judge the dryness or wetness of calcium chloride by its appearance; however, desiccants containing moisture indicators are commercially available. (Drierite® is anhydrous calcium sulfate impregnated with an indicator that is blue when dry and pink when wet.) A small amount of such a desiccant added to the anhydrous calcium chloride in a drying tube will indicate when the calcium chloride should be replaced.

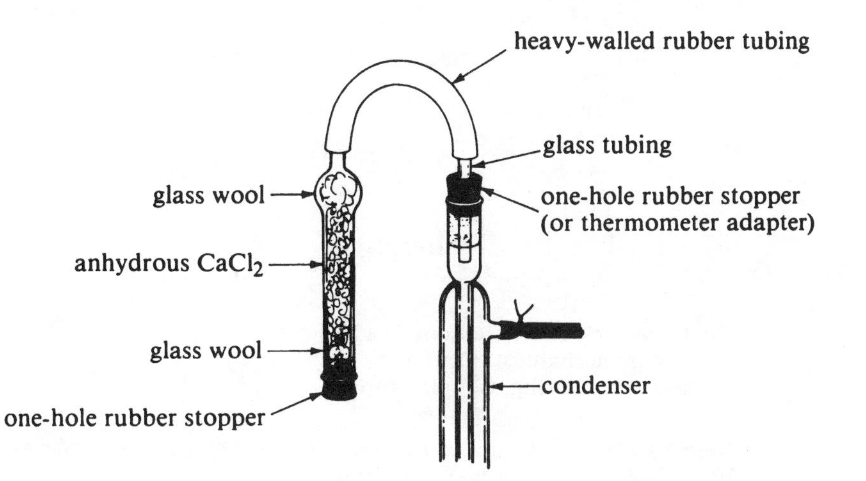

Figure 14.4 A straight drying tube connected to the top of a condenser.

14.6 Setting Up a Three-Neck Reaction Flask

Your laboratory may not be supplied with two- or three-neck flasks for your experiments; however, they are the standard reaction vessels in research and advanced organic laboratories. These flasks are convenient for reactions that require more than one attachment to the reaction flask.

Figure 14.5 shows a three-neck-flask arrangement that would be suitable for a Grignard reaction and many other types of organic reactions. Note the *mechanical stirrer,* which is set in the center joint. The stirrer column must be carefully aligned so that it does not bind. The *reflux condenser* is set into a side joint of the flask. The

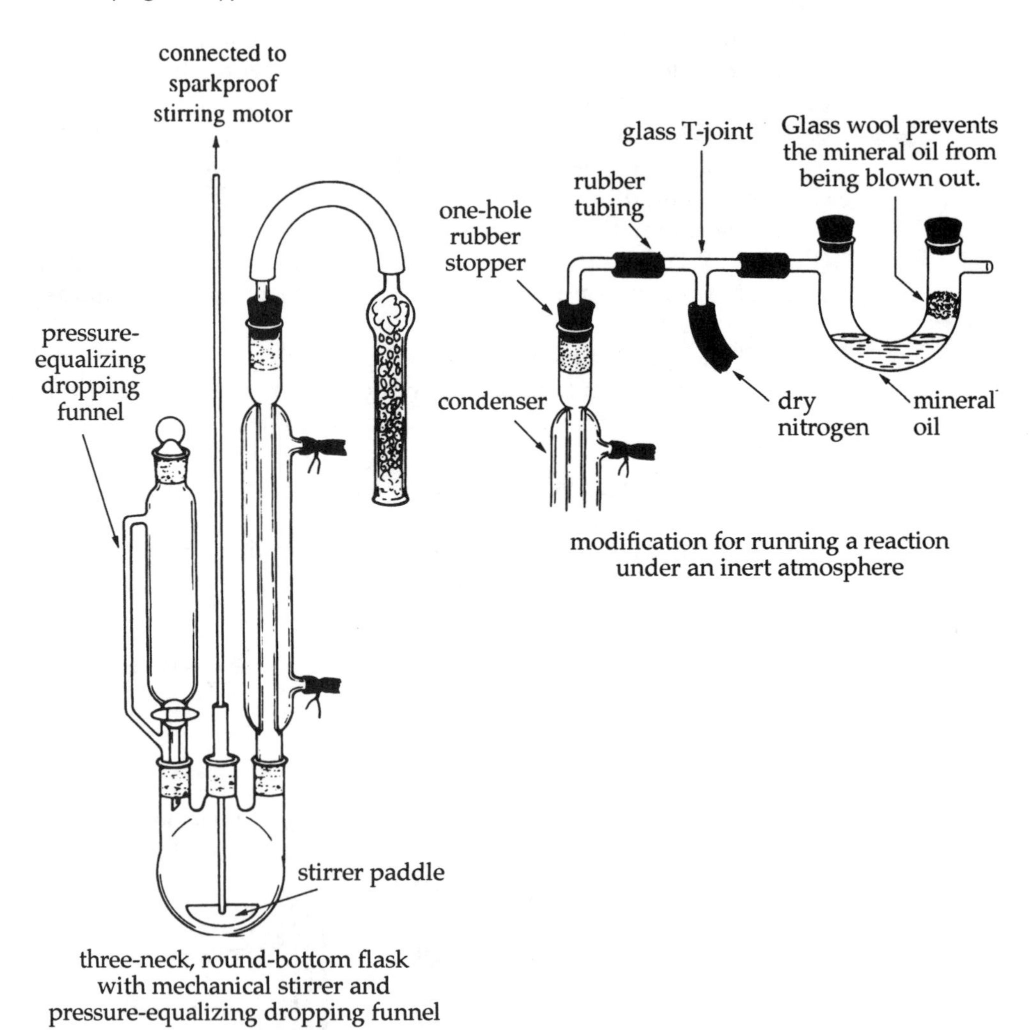

Figure 14.5 A three-neck-flask reaction assembly (clamps not shown). Be careful that solvent fumes do not flow over the stirring motor.

dropping funnel shown is a *pressure-equalizing dropping funnel.* Its sidearm is connected to the atmosphere within the flask (dry air and solvent fumes) so that its contents can be isolated from atmospheric moisture without pressurizing the funnel. An ordinary dropping funnel equipped with a drying tube on top would accomplish the same purpose.

Figure 14.5 also shows how the three-neck-flask apparatus can be modified for running a reaction under an inert atmosphere (dry nitrogen or argon). After the air is swept from the reaction apparatus, the inert gas is kept at a slightly positive pressure. The apparatus is not continuously swept with gas, because solvent vapors would be carried into the room. The mineral oil bubbler excludes air without sealing the system.

14.7 Hydrogen Halide Gas Traps

Reactions in which gaseous HCl and HBr are given off should be carried out in the fume hood. Unfortunately, many laboratories do not have sufficient hood space for an entire class. In such a situation, the HCl and HBr must be trapped.

Figure 14.6 shows a simple trap for a hydrogen halide. The beaker contains 5% aqueous NaOH (caustic, even at this dilution). The funnel is clamped so that it is *just above* the alkaline solution. (If the funnel is submerged, a pressure drop in the reaction vessel could cause the alkaline solution to be drawn into the reaction flask, with possibly disastrous consequences.)

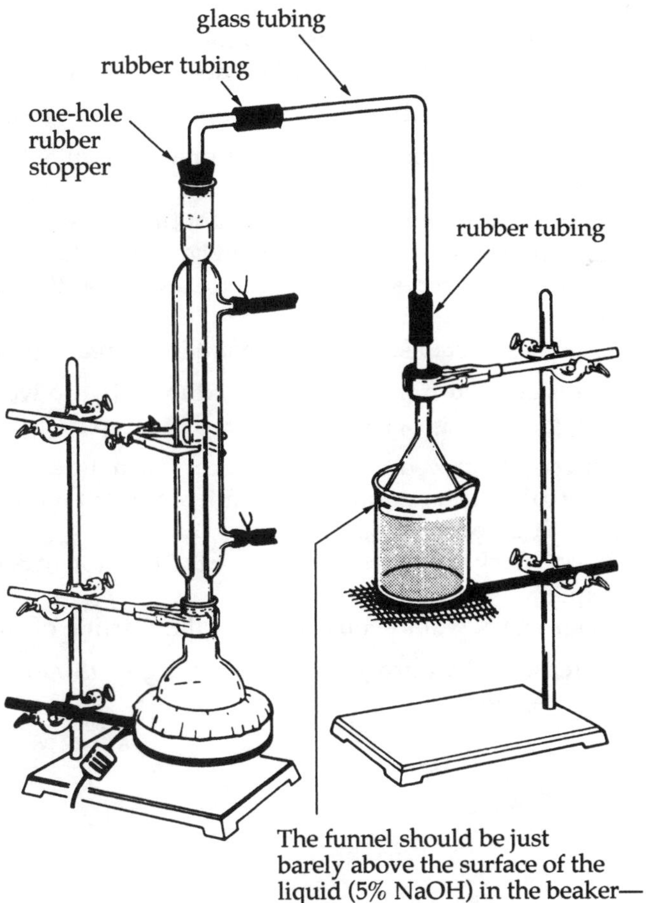
glass tubing
rubber tubing
one-hole rubber stopper
rubber tubing
The funnel should be just barely above the surface of the liquid (5% NaOH) in the beaker—it must not be submerged!

Figure 14.6 A gas trap for HBr or HCl.

Problems

14.1 Suggest reasons for the following procedures when you are setting up a reaction assembly.
(a) The ground-glass joints are lightly greased.
(b) Only a small amount of indicator desiccant is placed in a drying tube, instead of the tube being filled completely with this material.
(c) Powdered drying agents are not used in drying tubes.

14.2 List three ways of preventing a "runaway" reaction.

14.3 Why are the following procedures invalid?
(a) A boiling chip is not added to a reaction mixture that is to be boiled (but not stirred).
(b) Boiling chips are saved for use in a future experiment.
(c) After use, boiling chips are washed down the drain.

14.4 Why should a reaction never be carried out in a closed system?

14.5 Explain why the following are incorrect practices while heating a macroscale reaction under reflux.
(a) The water to the condenser jacket is not turned on.
(b) A gentle reflux is used rather than a vigorous reflux.
(c) The heat is adjusted so that the reflux level is at the very top of the condenser.
(d) The heat is adjusted so that the solvent in the reaction flask is not boiling.

14.6 When HCl is emitted from a reaction mixture, why do we not simply bubble the effluent gas through a solution of NaOH to trap it?

14.7 The experimental procedure in a chemical journal describes the exothermic reaction of 10.0 g of compound A in 25 mL of solvent with 75 mL of solution B in a 250-mL Erlenmeyer flask.
(a) What size flask would you choose for the reaction of 500 mg of compound A? Explain.
(b) What size flask would you choose for the reaction of 15.0 g of A? Explain.

14.8 Suggest a reason why nitrogen, rather than argon, is more commonly used as an inert atmosphere.

Infrared Spectroscopy

Infrared radiation (IR) is a portion of the electromagnetic spectrum between the visible and microwave regions. Infrared radiation is of interest to organic chemists because organic compounds absorb radiation in this region of the electromagnetic spectrum. The radiation increases the amplitudes of vibration of the bonds in the molecules; different types of bonds, and thus different functional groups, absorb infrared radiation of different wavelengths. An **infrared spectrum** is a plot of wavelength or frequency versus absorption. By inspecting an infrared spectrum obtained from a compound of unknown structure, a chemist can often identify the bond types and functional groups of the compound. Unfortunately, it is rarely possible to determine the complete structure of a compound by inspection of its infrared spectrum alone.

The theory of infrared spectroscopy is covered in your lecture text. Here, we will briefly describe infrared spectra and then, in more detail, interpretation of infrared spectra and the laboratory procedures used to obtain an infrared spectrum.

15.1 The Infrared Spectrum

Figure 15.1 shows the IR spectrum of 1-propanol. The lower horizontal scale of an infrared spectrum is given in a unit called **wavenumbers**, which have the unit *reciprocal centimeters* (1/cm, or cm^{-1}). Wavenumbers are referred to as *frequency*, and the spectrum has higher frequencies to the left as the typical spectrum is viewed. Alternatively, the lower scale of an infrared spectrum may be in **wavelengths** (λ), which have the unit micrometers (μm), where 1 $\mu m = 10^{-4}$ cm. The wavenumber may be calculated from the wavelength by the following equation:

$$\text{wavenumber in cm}^{-1} = \frac{1}{\lambda \text{ in cm}} = \frac{1}{\lambda \text{ in } \mu\text{m}} \times 10^4$$

The vertical scale of an infrared spectrum shows either **percent transmittance** (%T) or **absorbance** (A) of the radiation passing through the sample.

$$\%T = \left(\frac{\text{intensity}}{\text{original intensity}}\right) \qquad A = \log\left(\frac{\text{original intensity}}{\text{intensity}}\right)$$

A typical infrared spectrum plotted as % transmittance (%T) exhibits a "baseline" at the top that represents essentially no absorption of energy by the sample. At a wavelength at which the sample absorbs radiation, the tracing shows a decrease in %T. The absorption is thus recorded as a dip, called an *absorption peak* or *absorption*

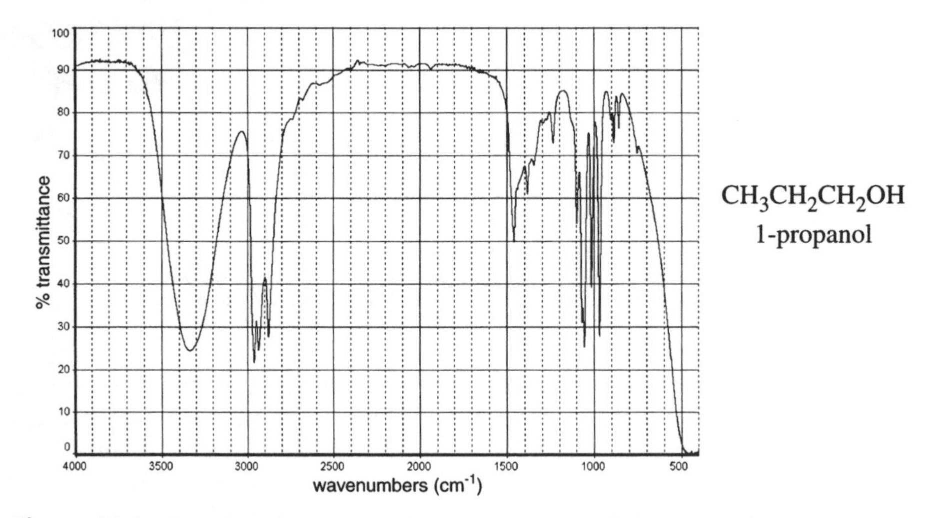

Figure 15.1 The infrared spectrum of 1-propanol, recorded on a Nicolet Impact 410 FT-IR.

band; the frequency or wavelength of the minimum point of an absorption band is used to identify that band.

15.2 Absorption of Infrared Radiation

Bonds within a molecule undergo a variety of **fundamental modes of vibration,** described by such terms as stretching, bending, scissoring, rocking, and wagging. Each type of vibration absorbs infrared radiation of its own characteristic wavelength, giving rise to a large number of peaks in an infrared spectrum. Each fundamental mode is also associated with *overtones,* or *harmonics,* similar to those produced by a guitar or other stringed instrument. These overtones add still more peaks to a spectrum.

The net result of all these phenomena is that an infrared spectrum contains a multitude of peaks, only a few of which are important in the correlation of the spectrum with an organic structure. Do not attempt to identify every small peak in an infrared spectrum! The key to interpreting an infrared spectrum is to inspect it for the presence or absence of only a few significant absorption bands.

Besides its position, the *relative intensity* of an absorption band is also useful to the organic chemist. The relative intensities depend partly on the relative numbers of particular groups within a molecule. For example, three CH_2 groups in a molecule absorb more radiation than one CH_2 group. The relative intensity of absorption by a group of atoms also depends on the *change in dipole* when energy is absorbed. A nonpolar grouping, such as C=C in the symmetrical alkene $(CH_3)_2C=C(CH_3)_2$, does not absorb infrared radiation. By contrast, the polar carbonyl group (C=O) shows strong infrared absorption.

15.3 Interpreting Infrared Spectra

Since the 1940s, when the infrared spectrophotometer became widely used, thousands of compounds have been subjected to this analytical technique. The result of these studies has been an empirical correlation of the positions of absorption of the various types of bonds and functional groups. This information is often summarized in charts called **correlation charts**. A typical correlation chart is shown inside the back cover of this book.

For purposes of organic structure identification, an infrared spectrum can be conveniently divided into two portions. The region from 1500–4000 cm^{-1}, to the left in the infrared spectrum, is especially useful for identification of the various functional groups. This region shows relatively strong absorption arising from stretching modes. The region to the right of about 1500 cm^{-1}, called the *fingerprint region*, is usually quite complex and often difficult to interpret; however, each organic compound has its own unique absorption pattern in this region. Figure 15.4 shows the spectra of two alkenes (which are geometric isomers); examine these spectra to see the differences in the fingerprint regions.

In beginning your study of infrared spectra, do not attempt to memorize all the frequencies of absorption of the various functional groups. Instead, learn to recognize the most important regions in the spectrum: the OH–NH region, the CH region, and the C=O region. In Figure 15.2, blocks indicate these general regions. When you are presented with the spectrum of an unknown compound, examine these portions of the spectrum first. After you have identified the important features, then check the spectrum against the tables and discussions that follow.

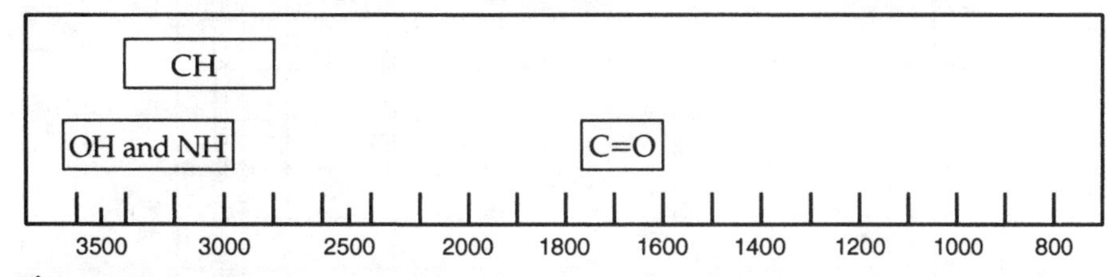

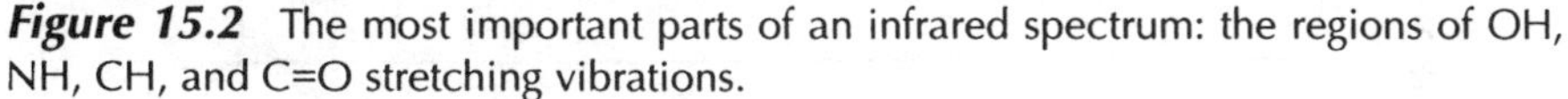

Figure 15.2 The most important parts of an infrared spectrum: the regions of OH, NH, CH, and C=O stretching vibrations.

15.4 Identifying Types of Compounds

A. Hydrocarbons

A distinguishing feature of the infrared spectra of hydrocarbons and other organic compounds containing CH bonds is the *CH stretching absorption*. **Alkanes** and **alkyl groups** (sp^3 CH) exhibit this absorption at 2800–3000 cm^{-1}. **Alkenes, alkynes,** and **aromatic compounds** exhibit CH stretching slightly to the left of this value, as Figure 15.3 and Table 15.1 both show.

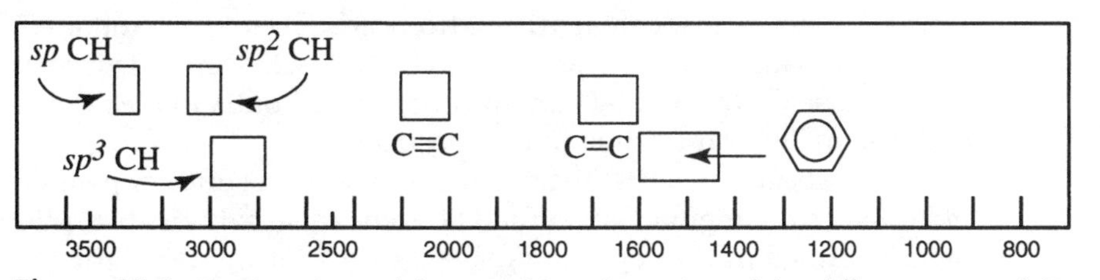

Figure 15.3 The locations of the stretching absorption of the different types of CH and C–C bonds.

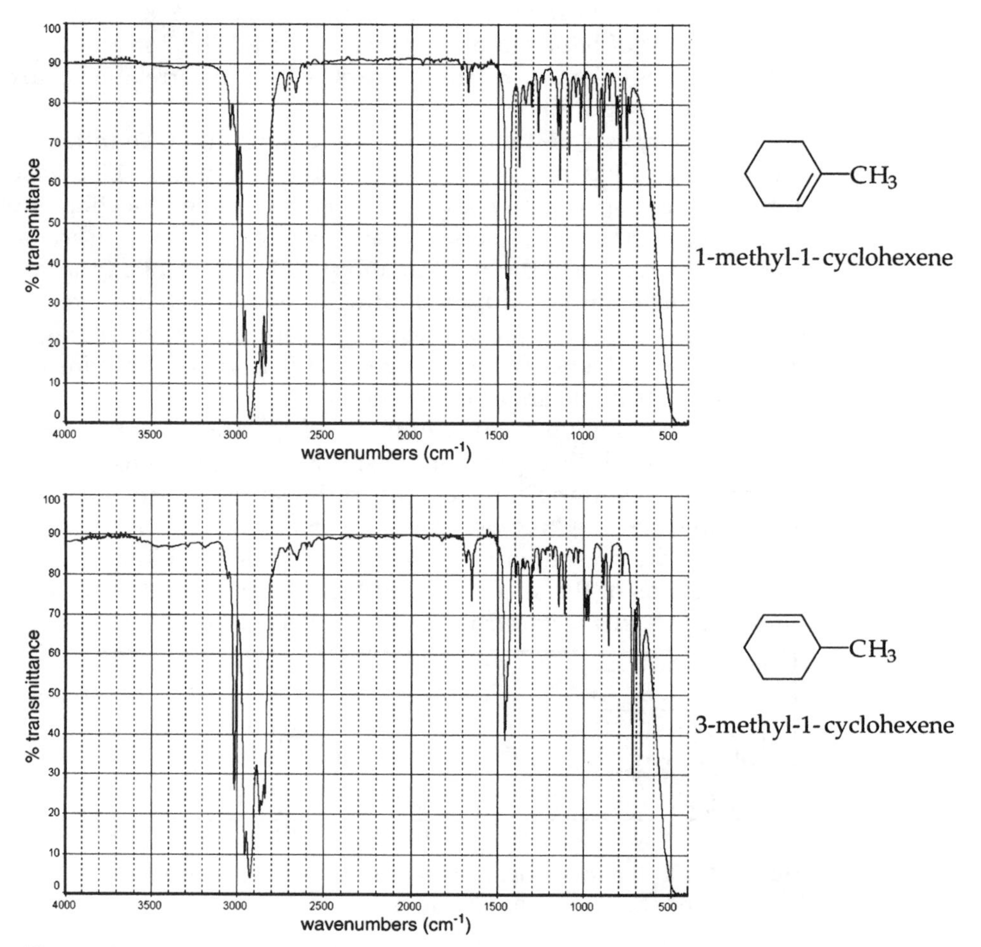

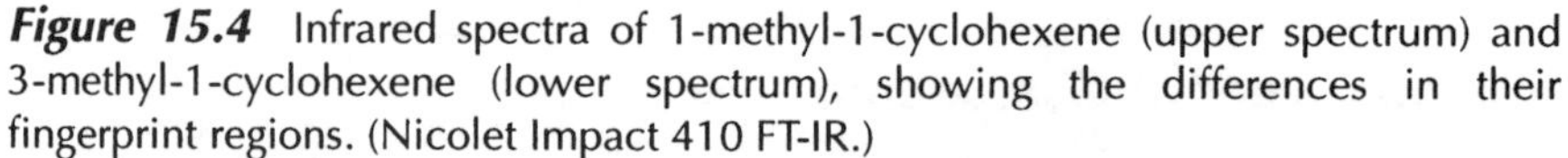

Figure 15.4 Infrared spectra of 1-methyl-1-cyclohexene (upper spectrum) and 3-methyl-1-cyclohexene (lower spectrum), showing the differences in their fingerprint regions. (Nicolet Impact 410 FT-IR.)

The C–C *stretching absorption* varies significantly, both in position and in intensity, for the various types of carbon-carbon bonds. The C–C stretching of alkanes and alkyl groups is generally too weak to be of value. **Alkenes** show generally weak, but often visible, C=C absorption at about 1600–1700 cm^{-1}. A highly unsymmetrical alkene shows stronger absorption than a more symmetrical alkene because a greater change in dipole results from its vibrations. Figure 15.5, the spectrum of 1-hexene, shows absorption arising from sp^3 CH, sp^2 CH, and C=C. Alkynes show a weak, but distinctive, C≡C absorption at 2100–2250 cm^{-1}. This absorption is distinctive because C≡N and Si–H are the only other bonds that absorb infrared radiation in this usually clear portion of the spectrum.

The spectra of **substituted benzenes** show a series of peaks (up to four) in the 1450–1600 cm^{-1} region. Substituted benzenes also exhibit absorption from CH bending vibrations in the region of 680–900 cm^{-1}, at the far right of a spectrum. This absorption often reveals the positions of substitution on the ring. Table 15.2 lists the

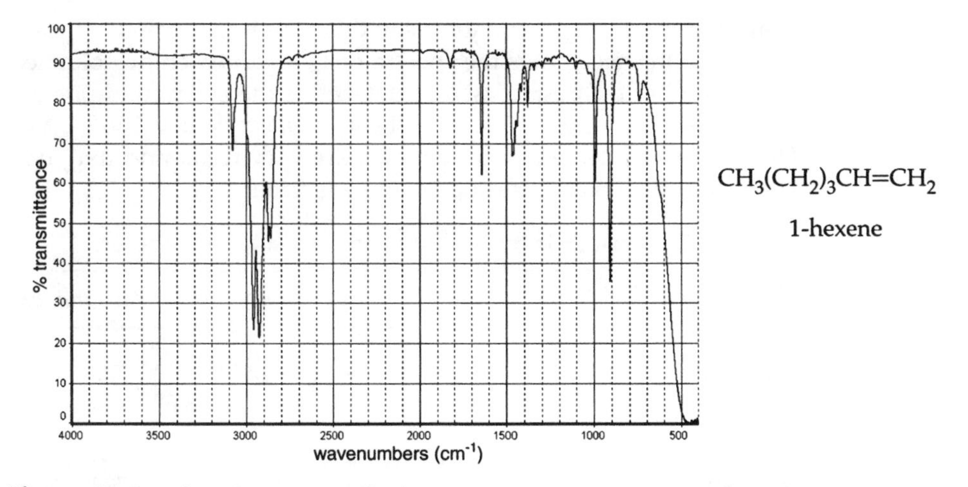

Figure 15.5 Infrared spectrum of 1-hexene, $CH_3(CH_2)_3CH{=}CH_2$, showing some types of C–C and C–H absorption. (Nicolet Impact 410 FT-IR.)

Table 15.1 C–C and C–H stretching absorptions.

Type of bond		*Absorption, cm^{-1}*
C–H:		
alkynyl	≡C–H	3300
aryl	C_6H_5–H	3000–3300
alkenyl	=C(–)H	3000–3100
alkyl	–C–H	2800–3000
aldehyde	O=C–H	2700–2900
C–C:		
alkynyl	C≡C	2100–2250
alkenyl	C=C	1600–1700
aryl	benzene ring	1450–1600

characteristic absorption patterns in this region for various types of substituted benzenes. Figure 15.6 shows the spectra of *o*-, *m*- and *p*-chlorotoluene.

Another technique used to determine the extent and positions of substitution on the benzene ring is to look at the *overtone pattern* at 1666–2000 cm^{-1} (Table 15.2). Unfortunately, these overtone bands are very weak and may not be visible unless the spectrum of a concentrated sample is taken.

Table 15.2 The C–H bending absorption of substituted benzenes.

Substitution		*Appearance*	*Absorption, cm^{-1}*
monosubstituted		two peaks	730–770 690–710
o-substituted		one peak	735–770
m-substituted		three peaks	860–900 750–810 680–725
p-substituted		one peak	800–860

B. Ethers, Alcohols, Phenols, and Amines

Ethers have a C–O stretching band that falls in the fingerprint region at 1050–1260 cm^{-1}. Because oxygen is electronegative, the stretching causes a large change in bond moment; therefore, the C–O absorption is often strong. Alcohols and other compounds containing C–O bonds also show absorption in this region. For example, 1-propanol shows C–O absorption at about 1075 cm^{-1} (see Figure 15.1).

The most conspicuous feature in the infrared spectrum of an **alcohol** or **phenol** is the OH absorption, to the *left* of CH absorption, at 3000–3700 cm^{-1}. (Again, refer to Figure 15.1.) The OH absorption in Figure 15.1 is of a *hydrogen-bonded OH group*. The spectrum of an alcohol in the vapor phase or in a dilute solution with a nonhydrogen-bonding solvent exhibits a sharper, weaker peak at about 3600 cm^{-1}, to the left of the broad, strong OH band usually observed for hydrogen-bonded alcohols. In some cases *two* OH peaks can be observed: one for the hydrogen-bonded OH and one for the nonhydrogen-bonded OH.

Amines containing NH bonds also show absorption to the left of CH absorption. If there are two hydrogens on an amine nitrogen ($-NH_2$), the NH absorption appears as a double peak. If there is only one H on the N, then one peak is observed (see Figure 15.7). Of course, if there is no NH bond (as in a tertiary amine, R_3N), then there is no absorption in the N–H stretching region. NH absorptions are weaker than O–H absorptions because oxygen is more electronegative than nitrogen; therefore, the O–H bond has a greater dipole change.

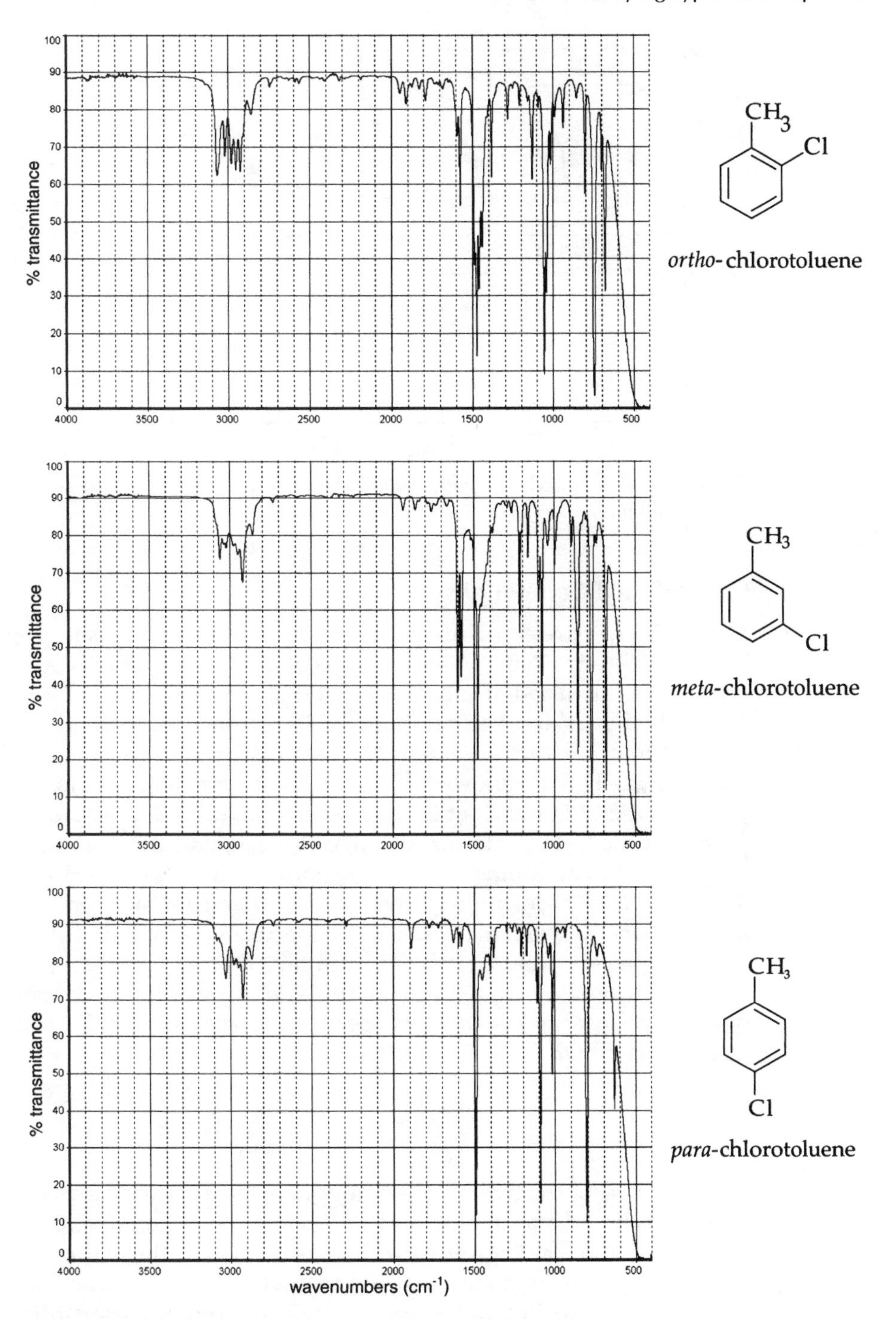

Figure 15.6 Infrared spectra for *o*-, *m*-, and *p*-chlorotoluene, showing the differences in absorption arising from C–H bending vibrations. (Nicolet Impact 410 FT-IR.)

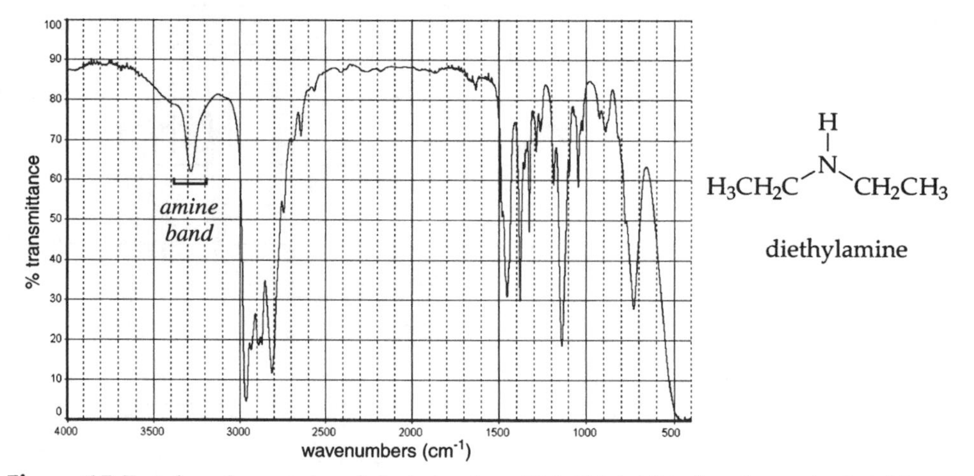

Figure 15.7 Infrared spectrum of diethylamine, $(CH_3CH_2)_2NH$, showing a single NH peak. (Nicolet Impact 410 FT-IR.)

C. Carbonyl Compounds

The carbonyl stretching mode gives rise to a strong, sharp peak somewhere between 1640–1760 cm^{-1}. The positions of C=O absorption for a variety of carbonyl compounds are listed in Table 15.3 (p. 173).

Aldehydes and **ketones** both exhibit strong carbonyl absorption. The important difference is that an aldehyde has an H bonded to the carbonyl carbon. This particular C–H bond shows two characteristic stretching bands (just to the right of the aliphatic C–H bond) at 2820–2900 cm^{-1} and 2700–2780 cm^{-1} (see Figure 15.8). Both of these C–H peaks are sharp, but weak, and the peak at 2900 cm^{-1} may be obscured by overlapping C–H absorption. The aldehyde C–H also has a very characteristic NMR absorption. If the infrared spectrum of a compound suggests that the structure is an aldehyde, the NMR spectrum (Technique 16) should be checked.

Carboxylic acids, either as pure liquids or in solution at concentrations in excess of about 0.01 *M*, exist primarily as hydrogen-bonded dimers rather than as discrete monomers. The infrared spectrum of a carboxylic acid is therefore the spectrum of the dimer.

```
     O---H—O
    //      \
R—C          C—R
    \       //
     O—H---O
```

Because of the hydrogen bonding, the O–H stretching absorption of carboxylic acids is broad and intense and slopes into the region of aliphatic carbon–hydrogen absorption (see Figure 15.9). The broadness of the carboxylic acid O–H band can often obscure both aliphatic and aromatic C–H absorption, as well as any other OH or NH absorption in the spectrum. The carbonyl absorption is of moderately strong intensity.

Esters exhibit both carbonyl absorption and C–O stretching absorption at 1100–1300 cm^{-1}. The C–O absorption may be used to distinguish esters from some

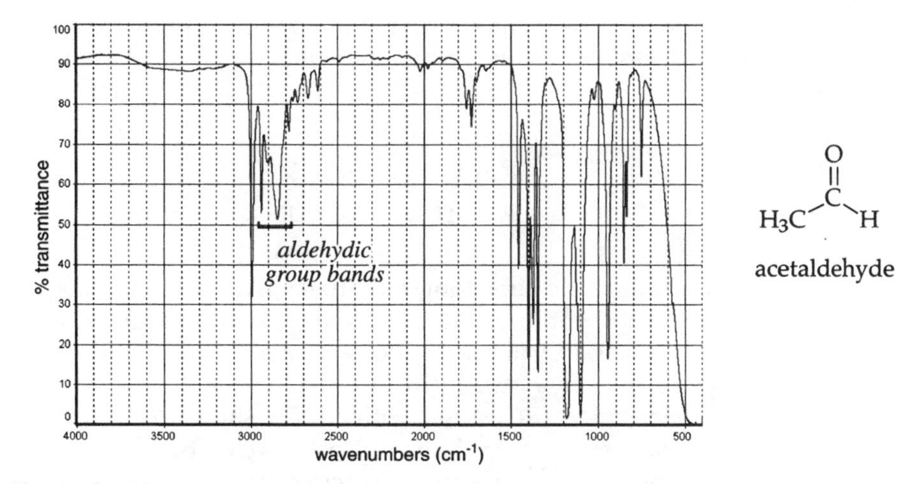

Figure 15.8 Infrared spectrum of acetaldehyde, CH_3CHO, showing the absorption by the aldehydic group. (Nicolet Impact 410 FT-IR.)

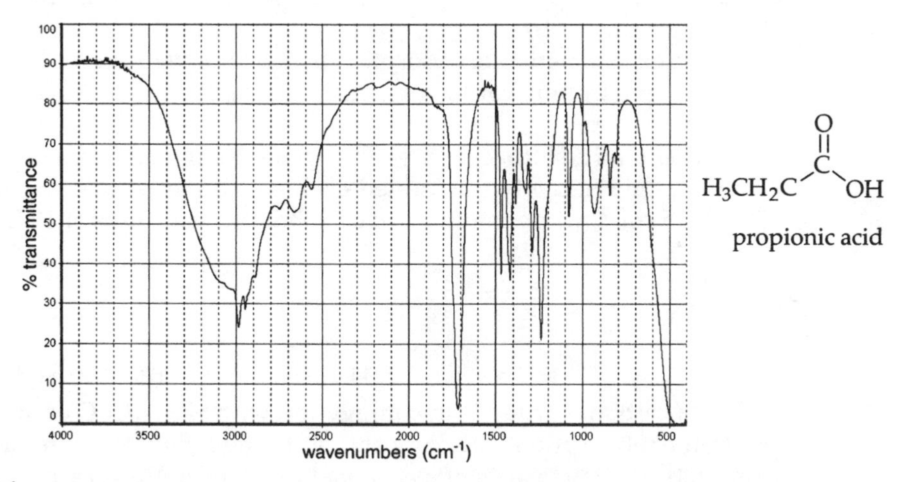

Figure 15.9 Infrared spectrum of propionic acid, $CH_3CH_2CO_2H$, showing the broad OH absorption. (Nicolet Impact 410 FT-IR.)

other carbonyl compounds. (Be careful: Carboxylic acids, anhydrides, and some other compounds also show C–O absorption.)

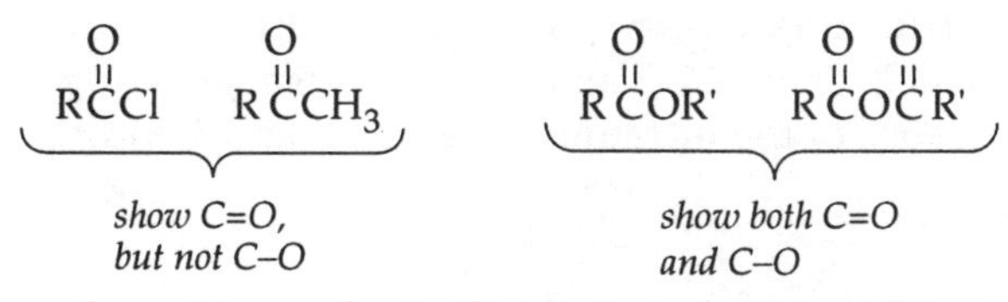

Amides may show *two* peaks in the carbonyl region. The *amide I band* arises from C=O. The *amide II band*, which appears between 1515–1670 cm^{-1}, just to the right of the C=O peak, arises from NH bending. Therefore, a disubstituted, or tertiary, amide does not show an amide II band.

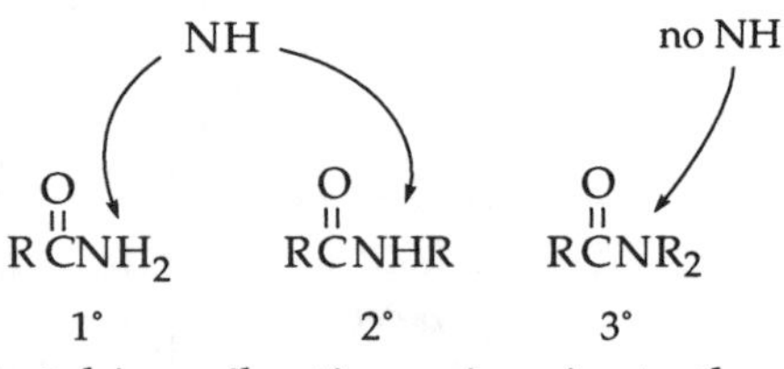

The amide NH stretching vibrations give rise to the absorption to the left of aliphatic CH absorption at 3125–3570 cm^{-1}. (This is about the same region where the NH of amines and OH absorb.) *Primary amides* ($RCONH_2$) show a double peak in this region. *Secondary amides* (RCONHR), with only one NH bond, show a single peak. *Tertiary amides* ($RCONR_2$), with no NH, show no absorption in this region.

The carbonyl infrared absorption of **acid chlorides** is observed at slightly higher frequencies than that of other acid derivatives. There is no other distinguishing feature in the infrared spectrum that signifies "This is an acid chloride."

A **carboxylic acid anhydride,** which has two C=O groups, generally exhibits a double carbonyl peak in the infrared spectrum. Anhydrides also exhibit C–O stretching around 1100 cm^{-1}.

The exact position of absorption by a carbonyl group in the infrared spectrum depends on a number of factors, such as ring strain, conjugation, and hydrogen bonding.

Ring strain shifts carbonyl absorption to higher frequencies (to the *left*). For example, the relatively unstrained cyclohexanones show carbonyl absorption at 1705–1725 cm^{-1}, while the strained cyclobutanone absorbs at 1775 cm^{-1}.

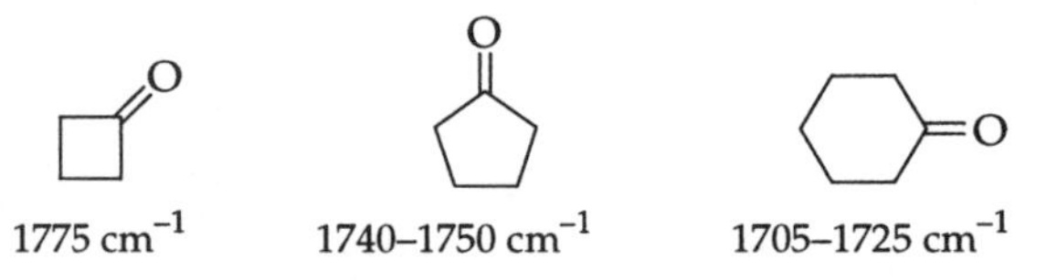

If a carbonyl group is in **conjugation** with either C=C or an aromatic ring, the position of absorption is usually shifted to the *right* (lower frequency) in the spectrum. Table 15.3 lists the shifts observed for some of these carbonyl compounds.

Hydrogen bonding of a carbonyl group with an NH or OH group results in a shift of the C=O absorption slightly to the *right* (lower frequency). This shift is observed in the spectra of carboxylic acids. Hydrogen bonding can also arise from a solvent such as chloroform (which can form weak hydrogen bonds). The effects of hydrogen bonding are also evident in the infrared spectrum of a β-diketone. A β-diketone exists partially in the *enol form* and shows OH absorption at 2500–2700 cm^{-1}. The carbonyl absorption of a β-diketone is broad and intense because of the intramolecular hydrogen bonding.

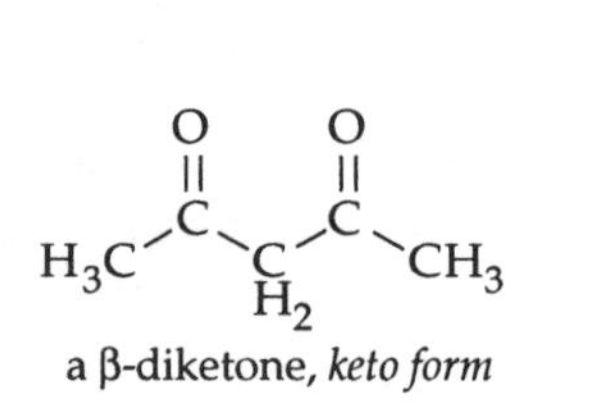

a β-diketone, *keto form*

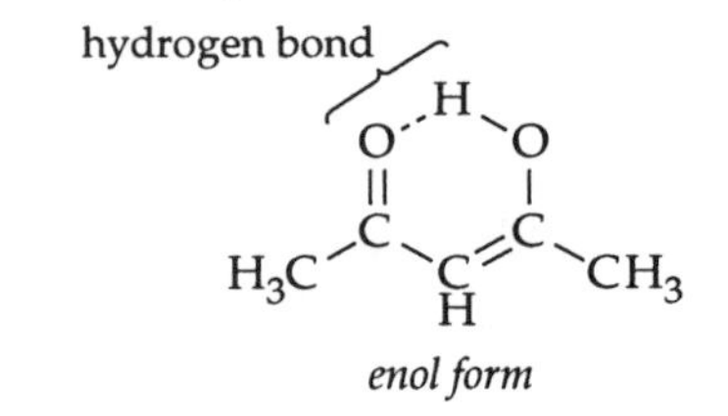

enol form

Table 15.3 Carbonyl stretching vibrations.

Type of compound	*Absorption, cm^{-1}*	*Comments*
aldehydes, RCHO	1720–1740	The aldehyde CH is distinctive.
conjugated	1685–1705	
ketones, $R_2C=O$	1705–1750	
conjugated	1660–1700	
β-diketones	1540–1640	OH absorption is also present.
carboxylic acids, RCO_2H	1700–1725	OH absorption is distinctive.
conjugated	1680–1700	
esters, RCO_2R	1740	Look for C–O absorption.
conjugated	1715–1730	
amides, $R(C=O)NR'_2$	1630–1700	1° and 2° amides also exhibit NH absorption.
acid chlorides, R(C=O)Cl	1785–1815	
conjugated	1770–1800	
acid anhydrides, R(C=O)O(C=O)R′	1740–1840	two carbonyl peaks
conjugated	1720–1820	
cyclic	1782–1865	

D. Other Types of Compounds

The stretching absorption of the C–X bond of an **alkyl halide** falls in the fingerprint region of the infrared spectrum, from 500–1430 cm^{-1}. For example, alkyl chlorides absorb at about 700–800 cm^{-1}. Without additional information, the presence or absence of a band in this region cannot be used for verifying the presence of a halogen in an organic compound.

Nitro compounds show two strong absorption bands in their infrared spectra: at 1500–1650 cm^{-1} and at 1250–1350 cm^{-1}. In addition, aromatic nitro compounds may show absorption at 750–850 cm^{-1}.

Nitriles (RC≡N) exhibit the same characteristic triple-bond absorption as do alkynes (2000–2300 cm^{-1}). The infrared spectrum alone cannot be used to distinguish a nitrile from an alkyne. However, a peak at 3300 cm^{-1} in the CH region would suggest the presence of a –C≡CH grouping.

15.5 Preparing a Sample for an Infrared Spectrum

A **sample cell** is a sample container that fits into the spectrophotometer in the path of the infrared radiation. Different types of cells are available for solids, liquids, solutions, and even gases. The "windows" of most common infrared cells, which allow the infrared radiation to pass through the sample, are polished sodium chloride plates. Sodium chloride is used because it is transparent to infrared radiation in the region of interest.

Sodium chloride plates and cells are relatively expensive because each plate is cut from a single, giant crystal and then polished. The plates and cells are also fragile and sensitive to moisture. Atmospheric moisture, as well as moisture in a sample, causes the polished plate surfaces to become pitted and fogged. Fogged plates scatter and reflect infrared radiation instead of transmitting it efficiently; poor-quality spectra are the result. Fogged plates may also retain traces of compounds from previous runs, giving rise to false peaks. If it is necessary to use fogged plates, be sure to clean them carefully, and then run a blank spectrum of the plates without sample before running the spectrum of your sample. By doing this, you can determine if your spectrum will contain extraneous peaks.

To prevent fogging, use only scrupulously dried samples for spectral work. Handle the plates *only by their edges*, never by their flat surfaces, because moisture from your fingers will leave fingerprints. NaCl plates and solution cells must be cleaned with a dry solvent such as CH_2Cl_2 after use and stored in a desiccator.

Fogged thin-film plates can be polished by rubbing them with a circular motion on an ethanol–water-saturated paper towel laid on a hard surface. Check with your instructor before you attempt to polish a plate!

A. Liquid Samples

Liquid samples are usually analyzed *neat* (meaning "pure" or "without solvent") as **thin-liquid films**. A drop or two of the liquid is sandwiched between two NaCl plates, and then the plates are mounted on a holder in the spectrophotometer. Figure 15.10 illustrates this technique.

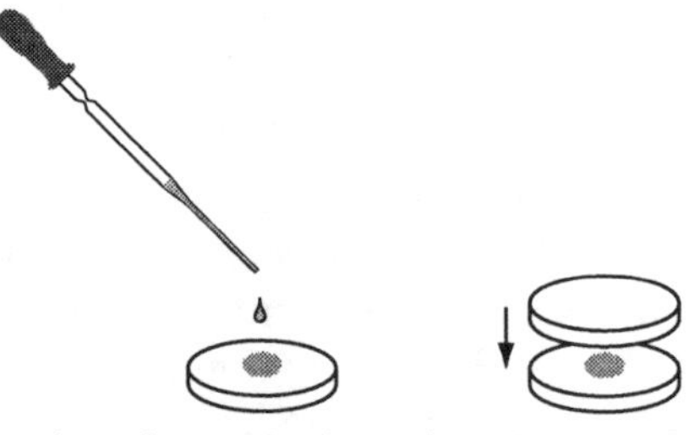

Figure 15.10 Preparing a liquid sample for infrared spectroscopy by the thin-film technique.

Few problems are encountered in thin-liquid film sampling, and these are easily solved. *Too much liquid* between the plates gives rise to a spectrum in which many of the peaks are too strong—in the 0–10% transmission range, or even off-scale. Also, leakage around the edges of the plates can contaminate the plate holder. To remove the excess liquid, wipe part of the sample off the plates (gently) with a dry tissue. Then rerun the spectrum.

Too little sample between the plates results in a spectrum with weak peaks. More sample should be added to the plates and the spectrum rerun.

After running a thin-liquid film spectrum, clean the NaCl plates gently with dry solvent and a tissue before returning them to the desiccator.

B. Solid Samples

(1) Thin-Solid Films

Most solid organic compounds can be run as a thin-solid film, similar to thin-liquid films discussed above. Although it does not give good quality spectra for all organic solids, this method is by far the easiest and quickest and should be tried first.

Dissolve a small amount of solid organic sample (5–10 mg) in a few drops of a volatile solvent. Suitable volatile solvents include methylene chloride, diethyl ether, and pentane. Place a drop of this solution on a single salt plate and allow the solvent to evaporate. You should observe a thin-solid film on the plate; if you do not, add another drop of the solution and allow it to evaporate. Mount the plate on the holder in the spectrophotometer. (There is no need to add another salt plate on top of the solid.)

Problems encountered with thin-solid films are similar to those with thin-liquid film sampling. *Too much solid* causes reduction of transmission to the 0–10% transmission range, with nondistinguishable peaks. To remove excess solid, remove the plate from the spectrophotometer and clean it. Dilute the solution of solid compound with solvent and repeat the sampling process.

If the peaks are weak, there is *too little solid* on the plate. Remove the salt plate, add more of the solution, allow the solvent to evaporate, and rerun the spectrum.

Solids with a low melting point can be run neat, like thin-liquid films. If you know that your solid melts near room temperature or even up to about 50°C, place a small amount of it in a test tube, warm the tube to melt the sample, then use a pipet to transfer a drop of the melted compound to a single salt plate. The sample should spread and solidify on the plate; it is then ready to be placed in the spectrophotometer.

(2) KBr Pellet

A **KBr pellet,** or **wafer,** is obtained by putting great pressure on a mixture of finely ground sample (dry!) and specially dried potassium bromide until a clear or translucent pellet is formed. Special die sets and either a hydraulic press, a hand-held press, or wrenches are used to press the pellet.

To prepare a KBr pellet, finely grind 1–2 mg of sample with 100 mg of anhydrous KBr (stored in the oven) in an agate or glass mortar. Pour about one-half of the sample mixture into a die set as directed by your instructor. Press the pellet under the supervision of your laboratory instructor. Place the die containing the pressed pellet in the spectrophotometer. When the spectrum is complete, remove the pellet from the die, wash and dry the die pieces, and store them in the oven.

(3) Mull

A **mull** is prepared by grinding the solid sample with an inert carrier (usually mineral oil) to the consistency of toothpaste or thick gravy. Different techniques can be used to grind the sample: an agate mortar, a ground-glass joint of a flask and a stopper, or two pieces of glass whose inner surfaces have been ground with carborundum. The ground mull is sandwiched between NaCl plates and run as a thin-liquid film.

A proper balance of mineral oil to sample is surprisingly difficult to achieve and is best determined experimentally. Begin by grinding together about 10–20 mg of solid with 2–3 drops of mineral oil. Use the spectrum to show if more sample or more mineral oil is needed. If the ratio of sample to mineral oil is correct, the strongest sample (not mineral oil) peaks will dip to about 40–50% transmission.

Mineral oil is a mixture of hydrocarbons; consequently, it shows hydrocarbon absorption bands in the infrared spectrum. Before running a mull spectrum, first run the spectrum of a thin film of mineral oil for reference. By doing this, you will be aware of which absorption bands in your sample spectrum arise from the carrier.

After running a mull spectrum, clean the NaCl plates gently with dry solvent and a tissue before returning them to the desiccator.

(4) Solution Spectrum

A solution spectrum is obtained by preparing a 5–10% (by weight) solution of the sample in a suitable solvent and placing the solution in an infrared solution cell. Different types of cells are available. By far the simplest cell is the cavity cell, which consists of a single block of sodium chloride with a 0.1-mm slit machined down the center of the block. These cells require only 0.05 mL of solution, which means that suitable spectra can be obtained with as little as 2–3 mg of sample. The cavity cell can be filled by delivering about 50 μL of a solution with a Pasteur pipet to the slit at the top of the cell.

The solvent chosen must dissolve the sample and must not contain any functional groups that would interfere with the spectrum of the sample. Typical infrared solvents are CCl_4 and $CHCl_3$. (Both of these solvents are toxic.) Of these, CCl_4 is the more useful because its only absorption is at 700–800 cm^{-1}, at the extreme right of the spectrum.

A solution spectrum shows peaks both of the sample of interest and of the solvent. Most modern infrared spectrophotometers are *Fourier transform* (FT-IR) instruments that incorporate data-handling programs. These programs have a feature in which the solvent spectrum (either freshly run or stored on the computer) can be subtracted from the solution spectrum, leaving only the peaks of the desired compound. If you are using an older-style instrument, you can place a matched cell filled with solvent alone in the reference beam of the spectrophotometer to compensate for solvent absorption bands. Or you can run the spectrum of a thin film of the solvent for reference.

After carrying out a solution spectrum, flush the cell thoroughly with the solvent. Dry the cell as instructed and return it to the desiccator.

15.6 Instrumentation

FT-IR instruments acquire a spectrum in a few seconds (or at most several minutes). Simply put the sample in the instrument and press the "scan" button as directed by your instructor.

FT-IR instruments have only one sample beam; older *dispersive* instruments have both a reference beam and a sample beam. A background spectrum is run periodically and stored in electronic memory. This background is used in place of the reference beam to "subtract out" interfering absorptions caused by molecules in the atmosphere of the sample chamber, poor-quality salt plates, or even solvents when a sample is run as a dilute solution or mull.

Once the spectrum is acquired, many manipulations can be made to appearance. You can mark the wavenumbers of your peaks, flatten a slanting baseline, enlarge the peaks to full scale, subtract out solvent peaks, expand an area to view details, and do many other operations. If the spectrum of a standard sample of the compound is available, you can superimpose the known spectra over the one you just acquired. Many computer programs have a search function so that you can search for matches of the spectrum of your compound in a database of known compounds. At the very least, you will want to print a copy of the spectrum to include with your lab report.

Additional Resources

A list of suggested readings which cover the theory of IR spectroscopy is published on the Brooks/Cole Web site. Other useful resources on the web site include references to compilations of IR spectra, links to tutorials, and practice problems. Please visit www.brookscole.com.

Problems

15.1 Arrange the following list of compounds in order of increasing intensity of the C=C stretching absorption (least-intense first), assuming identical molar concentrations in the sample cell:
(a) $CH_3CH{=}CHCH_2CH_3$
(b) $CH_2{=}CHCH_2CH_3$
(c) $CH_2{=}CCl_2$

15.2 The infrared spectrum of a student's sample shows a weak absorption band at 3710 cm^{-1}, yet the student is positive that the compound is not an alcohol or amine. Explain.

15.3 True or false? A double peak around 3300 cm^{-1} always indicates the presence of $-NH_2$. Explain your answer.

15.4 A student runs the infrared spectrum of cyclopentanone using chloroform as the solvent. The infrared spectrum shows a *double* carbonyl peak. Explain.

15.5 The infrared spectrum (thin-film) of a compound with the molecular formula C_7H_5N shows weak absorption at 3100 cm^{-1}, moderately strong absorption at 2230 cm^{-1}, and three peaks between 1400–1500 cm^{-1} as the principal absorption. Suggest a structure for this compound.

15.6 Tell how you would distinguish between each of the following pairs of compounds by their infrared spectra alone.

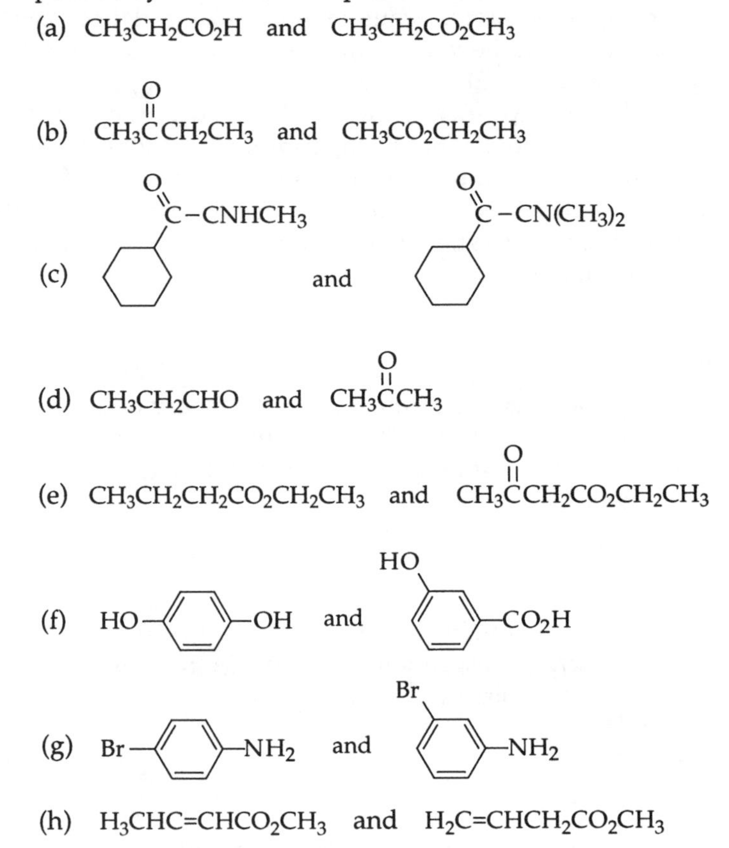

Technique 16

Proton Nuclear Magnetic Resonance Spectroscopy

Nuclear magnetic resonance (NMR) spectroscopy is based on the absorption of radio waves by certain nuclei in molecules exposed to a strong magnetic field. Proton NMR provides information about the molecular environment of hydrogen atoms in molecules. From this information you can often determine the number and relative locations of hydrogen atoms and how carbons and heteroatoms such as oxygen, nitrogen, and halides are bonded together to create the framework or connectivity of a molecule. Here we will briefly summarize the theory of NMR and the interpretation of proton NMR spectra. We suggest that you supplement your study of NMR spectroscopy by consulting your lecture text.

16.1 The NMR Spectrum

Protons, or hydrogen nuclei, in an organic molecule have spin and thus have small magnetic moments. When a sample is placed in an applied magnetic field (B_0), the magnetic moments of the protons become aligned either parallel or antiparallel to the field.

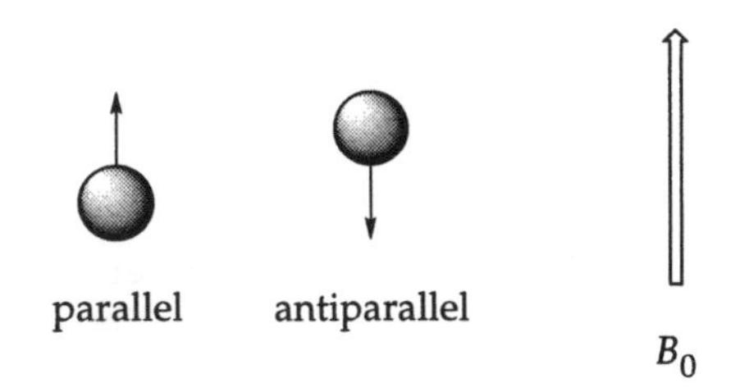

When energy of the proper frequency for a particular applied magnetic field is supplied to the sample, some of the protons in the parallel spin state absorb energy and "flip" to the higher-energy antiparallel spin state. In a NMR spectrometer, this absorption of energy is detected, amplified, and recorded.

The magnetic environments of different protons in a molecule are not all the same. An applied magnetic field acts on the electrons in a molecule to induce small magnetic fields within the molecule itself. These induced fields vary throughout the different parts of a molecule, depending on molecular structure. When electromagnetic energy, supplied by a radio-frequency oscillator, is applied to the molecules in the applied magnetic field, some protons absorb energy of slightly higher or lower

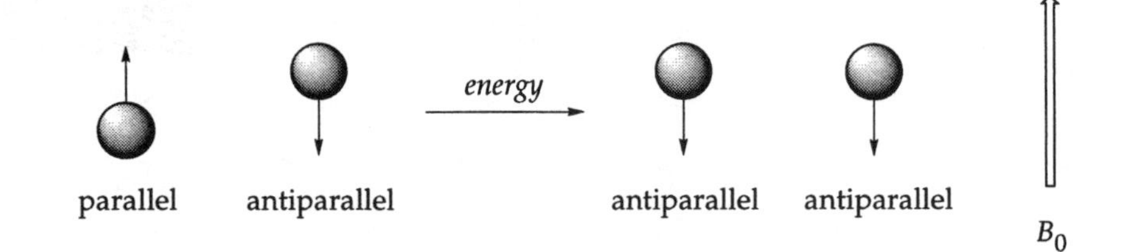

frequency than other protons at a particular applied B_0. NMR spectroscopy is based on this property.

Figure 16.1 is the NMR spectrum of methanol. The vertical scale is the **intensity of absorption**. Unlike in an infrared spectrum, the baseline is at the bottom and absorptions are observed as peaks, not dips.The horizontal scale is the position of absorption, or *ppm*: the **chemical shift**.

Chemical shift is recorded relative to the absorption of a standard, which is usually tetramethylsilane (TMS). Tetramethylsilane was chosen as a standard because the hydrogen nuclei of TMS absorb at a relatively high field or to the far *right* in the NMR spectrum.

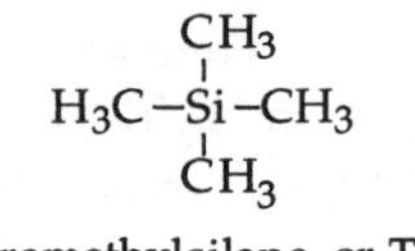

tetramethylsilane, or TMS

The absorption frequency of a particular proton varies with the spectrometer frequency; since different spectrometer frequencies are used at different facilities, the need arose for a convention for reporting chemical shift that was independent of operating frequency.* Thus by convention, chemical shift (δ) is defined as a measure of the difference between the absorption frequency of the proton of interest (ν_{sample} in Hz) and TMS (ν_{TMS} in Hz) divided by the operating frequency of the instrument (ν_o in MHz):

$$\delta = \frac{\nu_{sample} - \nu_{TMS}}{\nu_o} \times 10^6$$

Accordingly, chemical shift is reported as **ppm** (parts per million) downfield from TMS (which is 0 ppm by definition). For methanol (Figure 16.1), the chemical shifts of the two types of protons are reported as "2.1 ppm and 3.4 ppm downfield from TMS," or simply 2.1 and 3.4 ppm. Most hydrogen nuclei in organic molecules absorb between 0.0 and 10.0 ppm. Chemical shifts for a variety of types of protons are listed inside the back cover of this book.

* The spectrometer frequency is commonly 60, 300, 400, or 500 MHz.

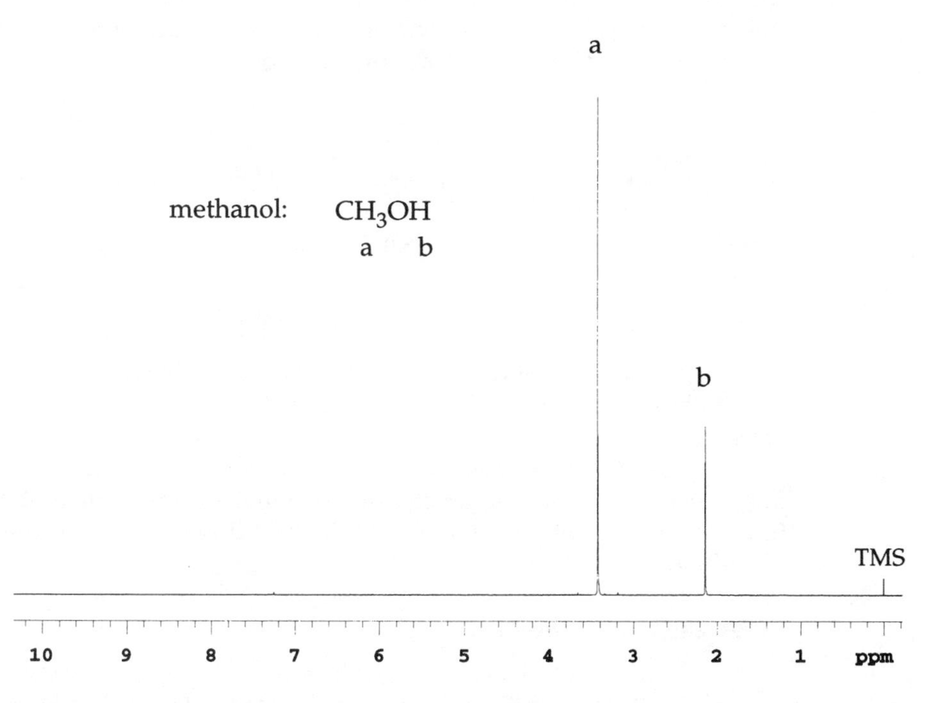

Figure 16.1 NMR spectrum of methanol. Solvent: $CDCl_3$. Instrument: Varian 300 MHz.

16.2 Interpreting NMR Spectra

A. Chemical Shifts

The inductive effect plays a role in the positions of absorption in NMR spectroscopy. A *more shielded* proton is one surrounded by a relatively greater electron density than another proton. A shielded proton absorbs *upfield* (to the right, closer to TMS, lower ppm) in the NMR spectrum. A less shielded, or *deshielded*, proton is one surrounded with a relatively lesser electron density and absorbs downfield (to the left, higher ppm) in the NMR spectrum.

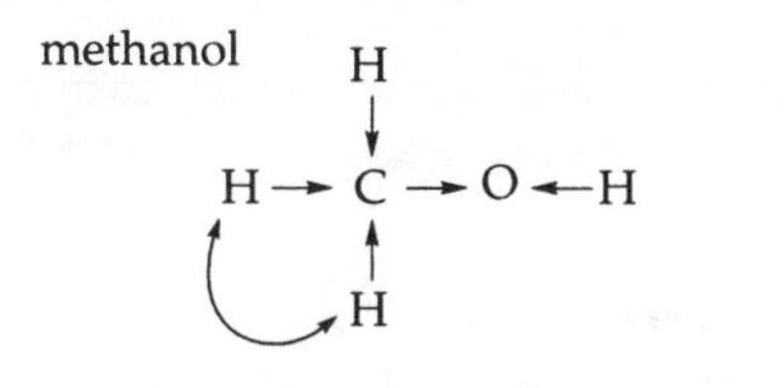

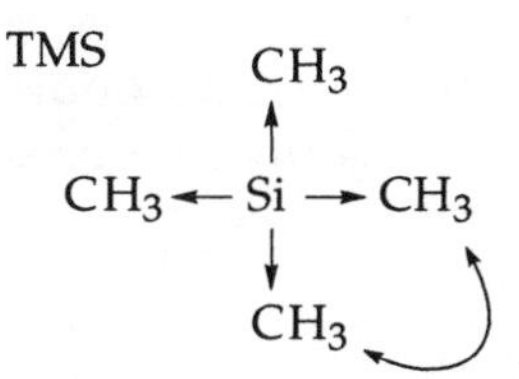

deshielded: absorb downfield because of electron withdrawal by electronegative oxygen atom

shielded: absorb upfield because of electron release by electropositive silicon atom

Protons in alkyl groups absorb from 0.8–1.5 ppm, depending on the substitution of the carbon. A proton on a carbon adjacent to a carbon bonded to oxygen absorbs further downfield than methyl groups because these protons are less shielded by the electron-withdrawing oxygen. Because of molecular magnetic fields induced in the pi bonds, protons attached to sp^2 carbons are deshielded and absorb from 4.9–5.9 ppm.

	RC(=O)OH	RC(=O)H	Ar–H	$R_2C{=}C(H)R$	R_2CHCl	RC(=O)CHR$_2$	$R_3C{-}H$
ppm:	10–12	9.4–10.4	6.0–8.0	4.9–5.9	3.5–3.7	2.0–2.7	0.8–1.5

→ *increasing shielding*

A proton in an aldehyde or carboxyl group absorbs far downfield because of the combination of the effects of the nearby pi bond and the electron-withdrawing oxygen.

Figure 16.2 illustrates the NMR spectra of representative compounds containing ester, alkyl halide, and aldehyde functional groups. Note that the aldehydic proton in propionaldehyde absorbs the farthest downfield of all the types of protons illustrated.

B. Areas Under the Peaks

Protons that are in the same magnetic environment in a molecule have the same chemical shift in a NMR spectrum. Such protons are said to be *magnetically equivalent protons*. Protons that are in different magnetic environments have different chemical shifts and are said to be *nonequivalent protons*. Equivalent protons in NMR spectroscopy are generally the same as chemically equivalent protons.

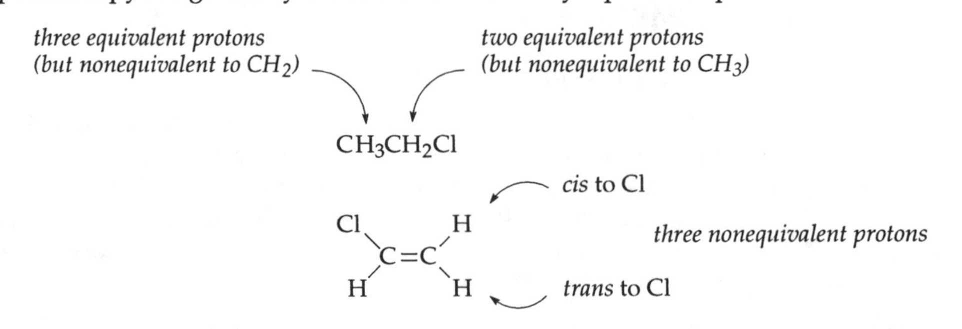

The **areas under the signals** in a NMR spectrum are proportional to the number of equivalent protons giving rise to that signal.

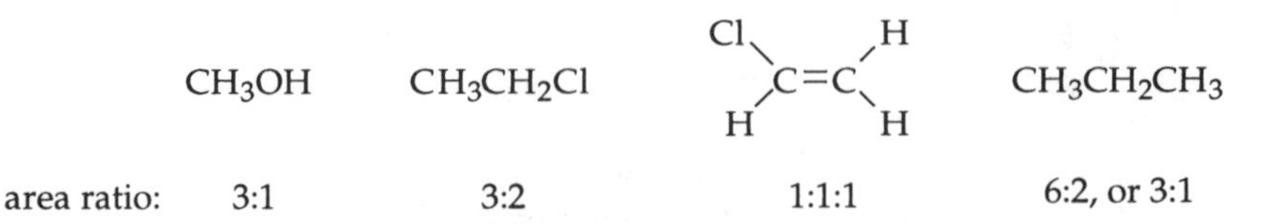

Integrators incorporated in NMR spectrometers measure the area under each peak and plot this value first as a stair-step line over the peaks, as illustrated in the spectrum of ethyl acetate, Figure 16.3. (This spectrum of ethyl acetate has also been expanded to show the details of the peaks. The full spectrum of ethyl acetate is

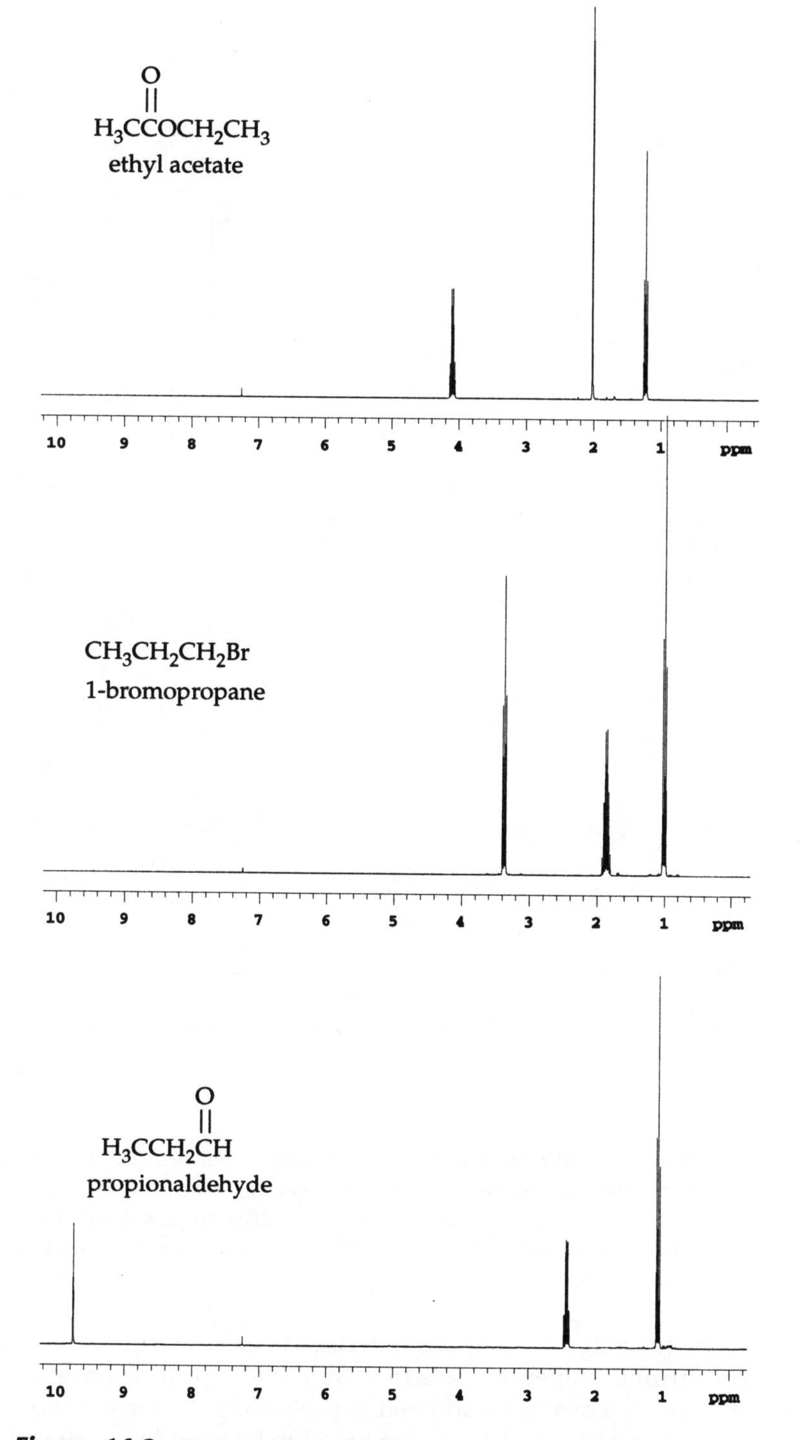

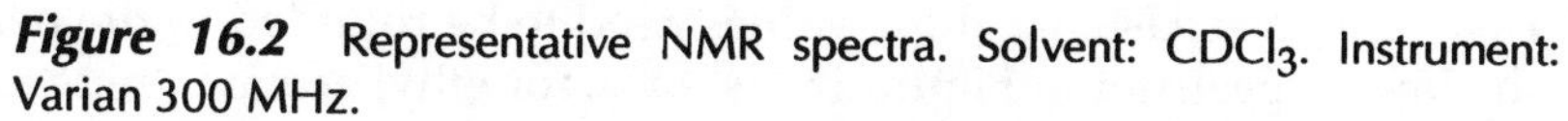

Figure 16.2 Representative NMR spectra. Solvent: $CDCl_3$. Instrument: Varian 300 MHz.

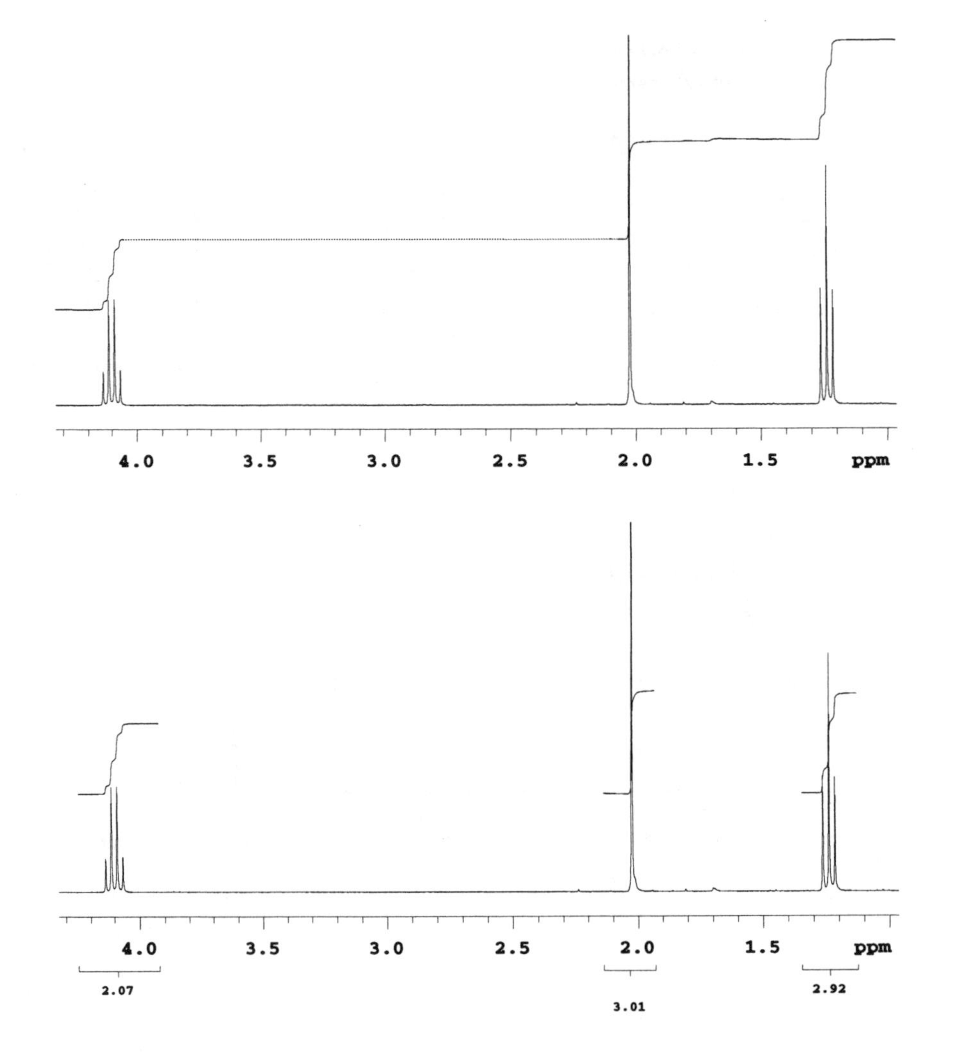

Figure 16.3 The NMR spectrum of ethyl acetate, expanded. The top spectrum illustrates the "stair-step" line that the instrument draws over the peaks. The lower spectrum illustrates the spectrum after the operator has directed the instrument to calculate the areas under the peaks. The integration values are printed below the ppm scale. Solvent: $CDCl_3$. Instrument: Varian 300 MHz.

shown in Figure 16.2.) Computer-interfaced NMR instruments, under the direction of the instrument operator, are then able to determine a numerical value of the height of the step over each absorption peak and print the value on the spectrum for easy reference. Usually these values need to be rounded to the nearest integer. As the lower spectrum in Figure 16.3 shows, for ethyl acetate, there are three types of protons: two of one kind at 4.1–4.3 ppm; three of another at ~2ppm; and three of still another type at 1.4–1.5 ppm.

C. Spin-Spin Splitting

The NMR signal for a proton (or group of equivalent protons) is split if the proton "sees" neighboring protons on adjacent carbons that are nonequivalent to it.

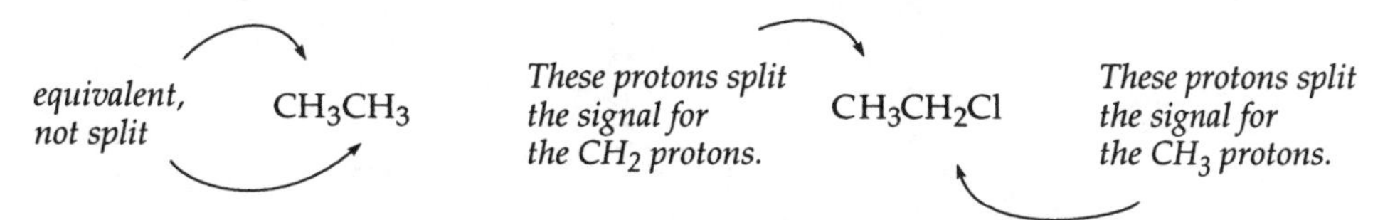

The splitting of signals is called **spin–spin splitting**, and protons splitting the signals of one another are called **coupled protons**. Spin-spin splitting arises from the differences in the effective magnetic field due to parallel and antiparallel spin states of the neighboring protons. Protons that have the same chemical shifts do not split each other's signals. Only neighboring protons that have *different chemical shifts* cause observable splitting.

For many compounds, we can predict the number of spin-spin splitting peaks in the NMR absorption of a particular proton (or a group of equivalent protons) by counting the number (n) of neighboring protons nonequivalent to the proton in question and adding 1. This is called the $\boldsymbol{n+1}$ **rule**. Some different types of splitting patterns that follow the $n + 1$ rule are listed in Table 16.1 (p. 187).

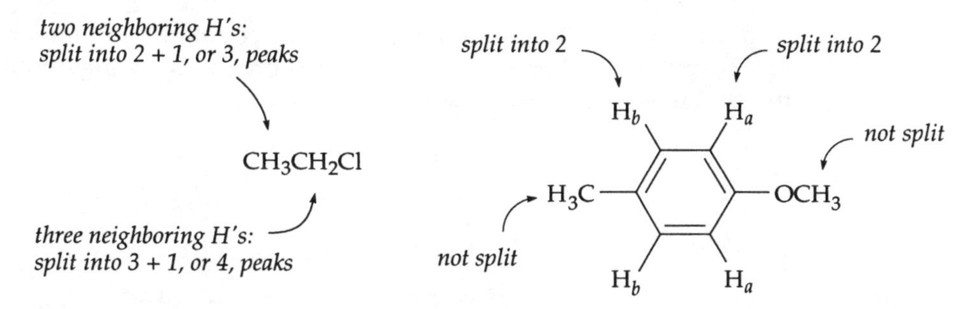

Assigning the peaks in a NMR spectrum is a matter of putting together the pieces of the puzzle: the chemical shifts of the different types of protons; the number of the different types of protons; and the splitting patterns. Figure 16.4 shows the interpreted NMR spectrum of ethyl acetate. The three protons at ~2 ppm are unsplit and somewhat deshielded, corresponding to the $-CH_3$ group adjacent to the carbonyl group. The three protons at 1.4–1.5 ppm are the most shielded since they are the farthest from the carbonyl, and they are split into three peaks by the two neighboring $-CH_2-$ protons. The two protons at 4.1–4.2 ppm are the most deshielded (a combination of the carbonyl and other effects) and are split into four peaks by the neighboring three protons on the $-CH_3$ group.

D. Coupling Constants

The separation between any two peaks of a split signal is called the **coupling constant** $\boldsymbol{J}$ and varies with the environment of the protons and their geometric relation-

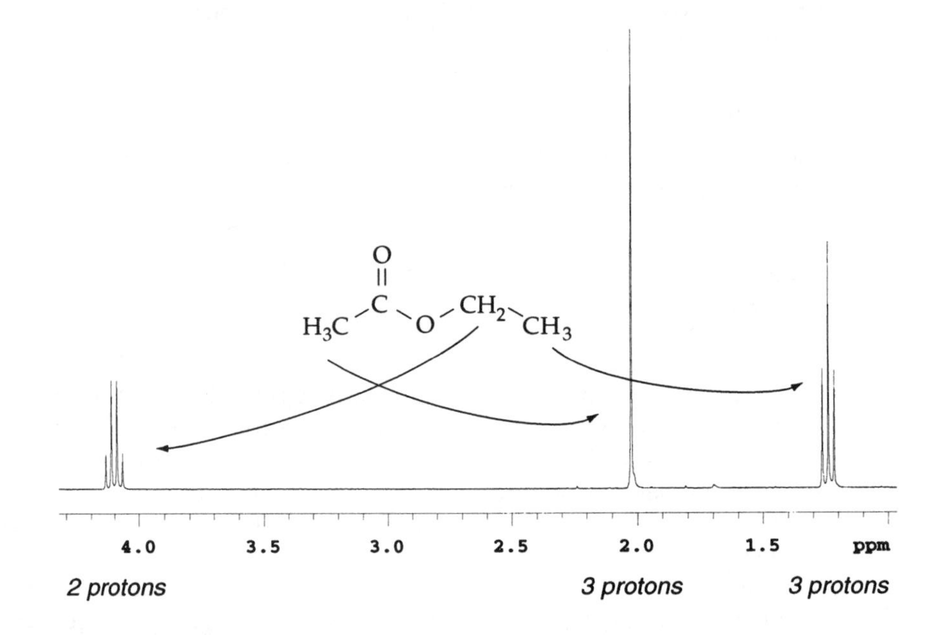

Figure 16.4 The expanded NMR spectrum of ethyl acetate, showing the splitting pattern and the assignment of the peak absorptions to the different types of protons in the molecule. Solvent: $CDCl_3$. Instrument: Varian 300 MHz.

ship to each other. The symbol J_{ab} means the coupling constant for H_a split by H_b or for H_b split by H_a. For any pair of coupled protons showing doublets, the J value is the *same* in each of the doublets.

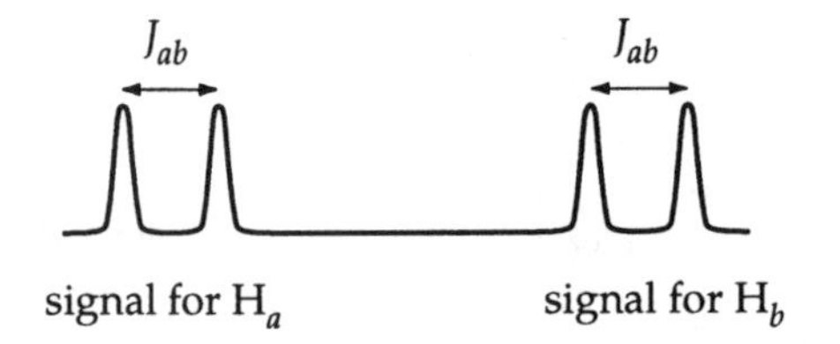

If the signal for a proton is split into a triplet, the distance between two adjacent peaks in the triplet is J_{ab}; thus, the width of the triplet is $2J_{ab}$. Similarly, the width of a quartet is $3J_{ab}$, and so forth.

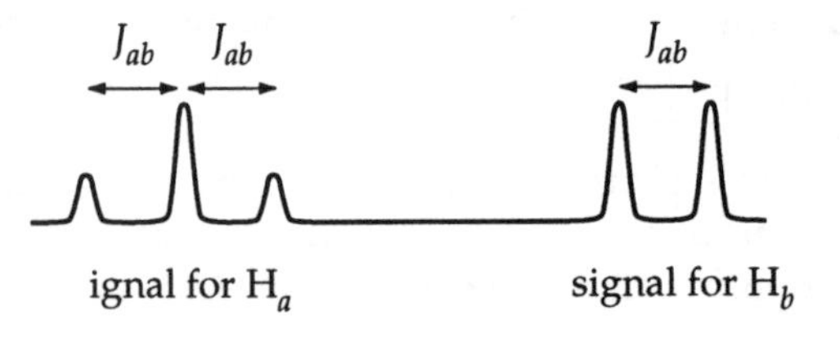

The units used for J values are Hz (cycles per second). The magnitude of the coupling constant is calculated by multiplying the separation of the lines in δ units (ppm) by the resonance frequency of the spectrometer in MHz. Table 16.2 lists some typical coupling constants.

Table 16.1 Some simple spin-spin splitting patterns in NMR spectra.

Partial structure[*]	*Number of neighboring nonequivalent H's*	*Signal appearance*	*Signal name*
$-\overset{\mid}{\underset{\mid}{\mathrm{C}}}\underline{\mathrm{H}}$	0	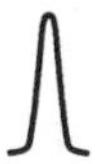	singlet
$>\mathrm{CH}\overset{\mid}{\underset{\mid}{\mathrm{C}}}\underline{\mathrm{H}}$	1	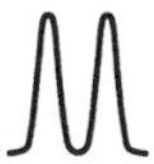	doublet
$-\mathrm{CH_2}\overset{\mid}{\underset{\mid}{\mathrm{C}}}\underline{\mathrm{H}}$	2	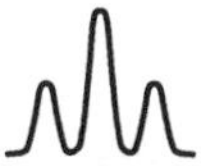	triplet
$\mathrm{CH_3}\overset{\mid}{\underset{\mid}{\mathrm{C}}}\underline{\mathrm{H}}$	3	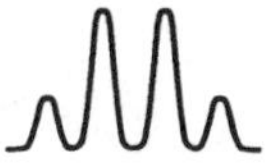	quartet
$(\mathrm{CH_3})_2\mathrm{C}\langle^{\diagup}_{\underline{\mathrm{H}}}$	6	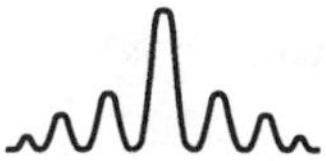	septet

* In each example, only the signal for the underlined proton is shown.

Coupling constants allow us to determine which protons are coupled with one another in a complex spectrum. If the J values are the same, the protons may be coupled. If the J values are different, the protons in question are not splitting the signals of one another, but are being split by some other proton.

E. Complexities in NMR Spectra

(1) Leaning

If the chemical shifts of coupled protons are fairly close, unsymmetrical splitting patterns (such as doublets or triplets) are observed. The signals seem to "lean" toward each other, with the inner peaks larger than the outer peaks. Leaning is a function of the differences in chemical shifts—the closer the chemical shifts, the greater the leaning.

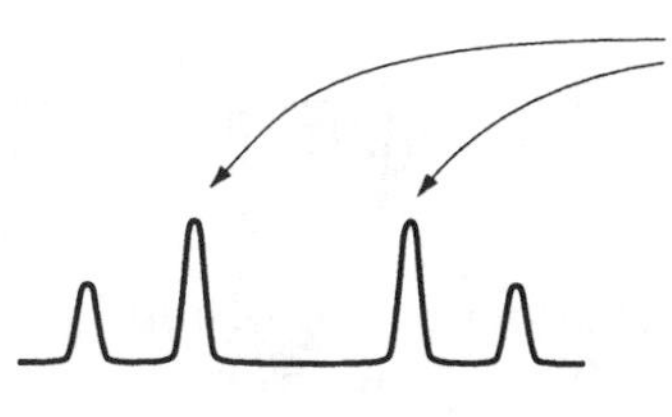

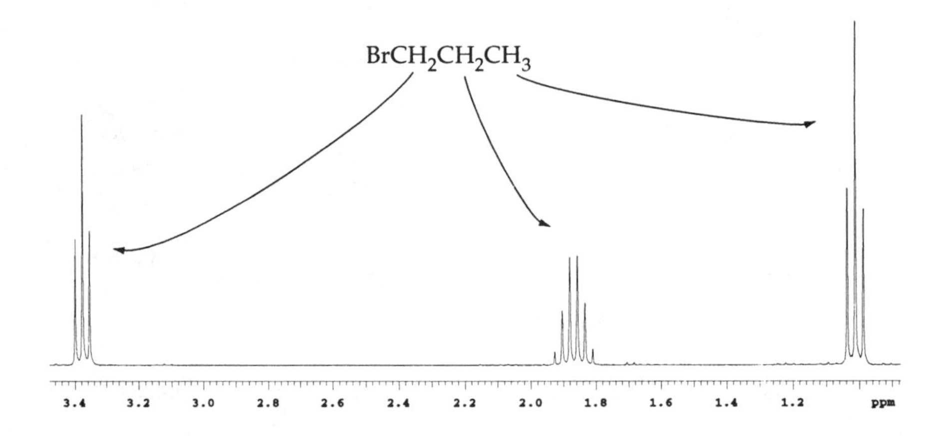

Figure 16.5 The expanded NMR spectrum of 1-bromopropane. Solvent: $CDCl_3$. Instrument: Varian 300 MHz.

The expanded NMR spectrum of 1-bromopropane illustrates the phenomenon of leaning (Figure 16.5). Note how the triplet at 1 ppm "leans" towards the sextet at 1.8–2.0 ppm.

(2) Protons Split by More Than One Type of Neighboring Proton

Protons split by more than one type of neighboring proton usually do not obey the $n + 1$ rule. Consider a case in which a proton is split by two other types of protons with different J values.

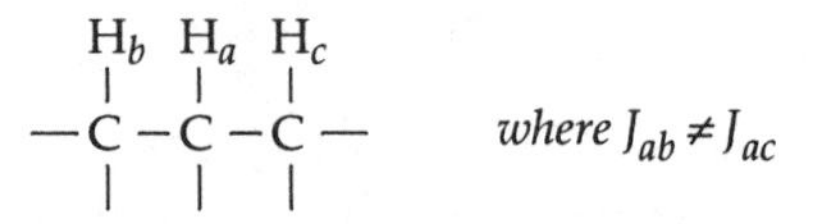

The $n + 1$ rule is obeyed only if J_{ab} and J_{ac} are the same, or very nearly the same. When these two coupling constants are different, the signal for H_a is split by H_b into two peaks and by H_c into two more peaks, so that four peaks, all with equal areas, result.

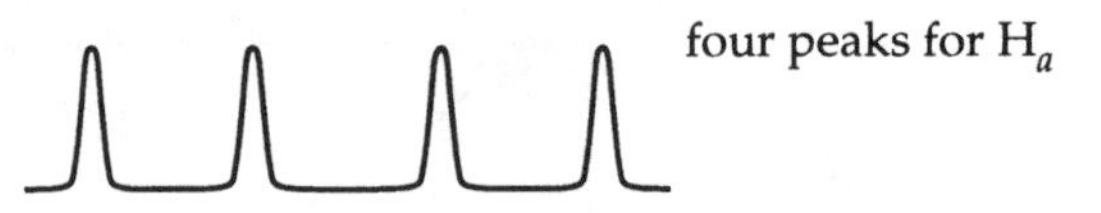

The NMR spectrum of styrene is shown in Figure 16.6. The 3 alkenyl (vinyl) protons give rise to 12 peaks (the 4 peaks from 5.2–5.8 ppm are each actually double

Table 16.2 Selected coupling constants.

Partial structure	*Coupling constant, J_{ab} (Hz)*
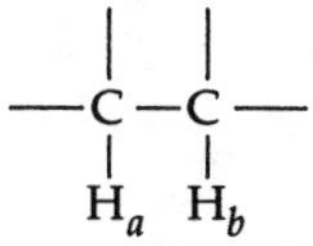 —C—C— (H_a, H_b)	6–8
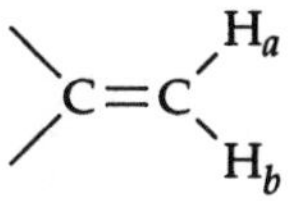 C=C (H_a, H_b)	0–3.5
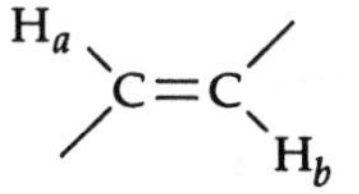 H_a C=C H_b	11–18
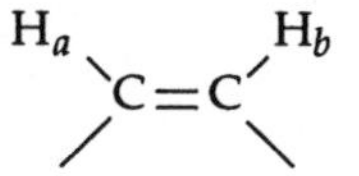 H_a H_b C=C	6–14
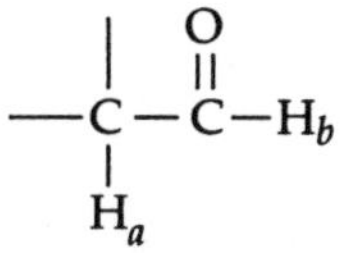 —C—C(=O)—H_b (H_a)	1–3
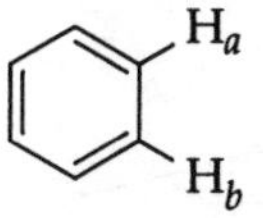 H_a H_b (benzene ring)	7–10
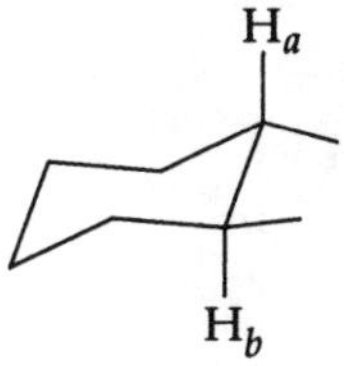 H_a H_b (cyclohexane ring)	10–13
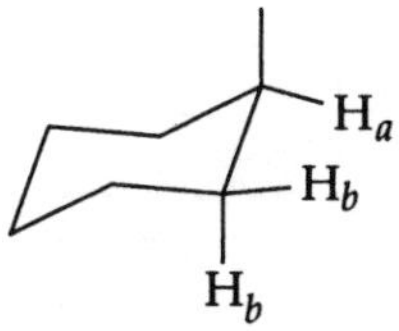 H_a H_b H_b (cyclohexane ring)	*e, e*: 3–5 *a, e*: 3–5

peaks, although it may be hard to see in the illustration). Eight to twelve peaks are typical of $-CH{=}CH_2$ groups. The 12 peaks result from the signal for each of the 3 alkenyl protons being split into 4 peaks by 2 types of neighboring protons.

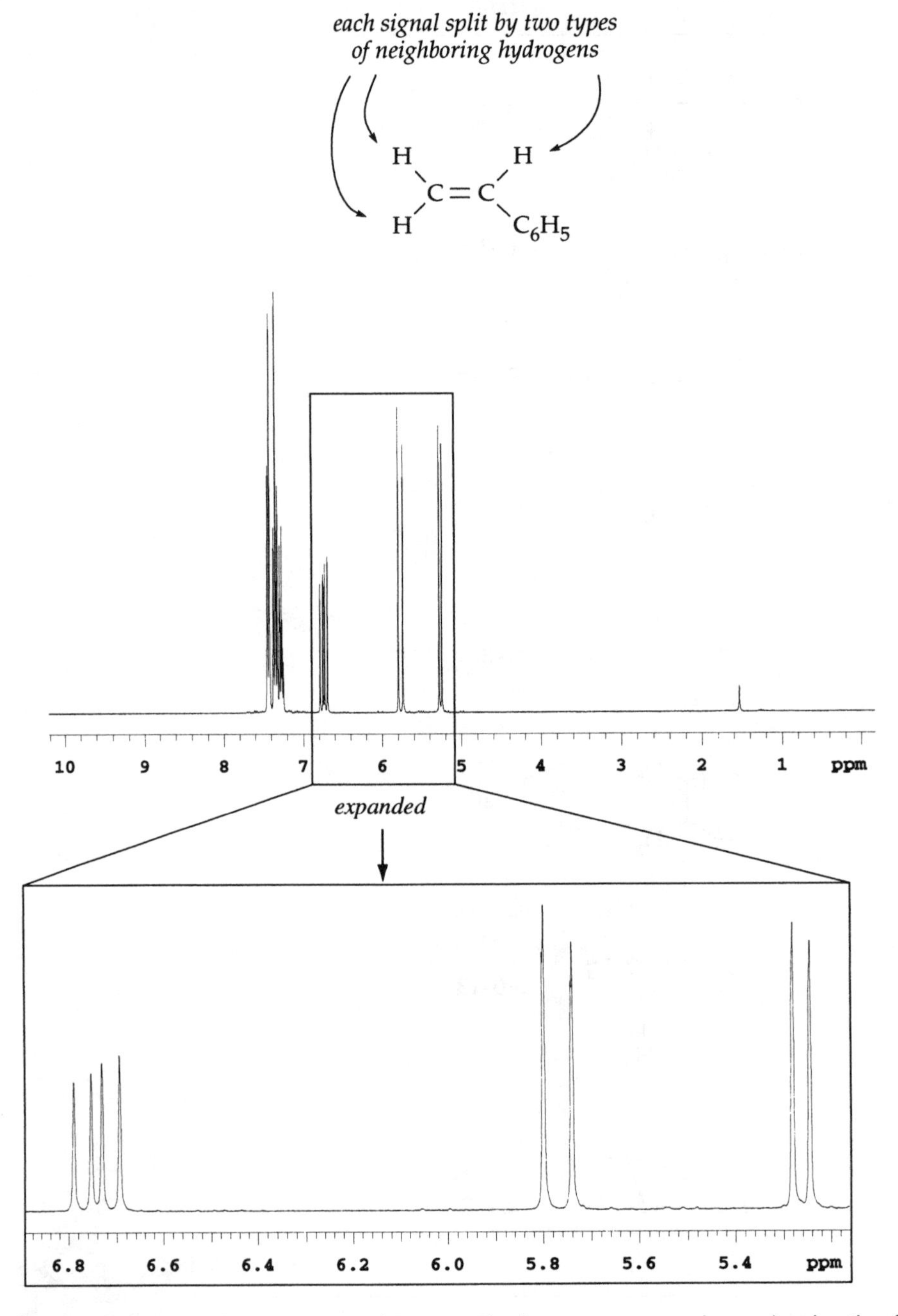

Figure 16.6 The NMR spectrum of styrene. The lower spectrum shows the details of the splitting of the alkenyl protons. Solvent: $CDCl_3$. Instrument: Varian 300 MHz.

(3) Chemical Exchange

On the basis of the preceding discussions, one would expect the NMR spectrum of methanol to show a doublet (for the CH_3 protons) and a quartet (for the OH proton). If the spectrum of methanol is determined at a very low temperature (–40°) or in specially prepared solvents such as CCl_4 stored over Na_2CO_3, this splitting is observed. However, if the spectrum is run at room temperature in ordinary solvents, only two singlets are observed, as can be seen in the spectrum of methanol (Figure 16.1).

The reason for this behavior of methanol is that alcohol molecules (and also amines) undergo rapid reaction with each other in the presence of a trace of acid, exchanging OH (or NH) protons in a process called **chemical exchange**. This exchange is so rapid that neighboring protons cannot distinguish any differences in spin states.

$$CH_3OH + H'^{+} \rightleftharpoons CH_3\overset{+}{O}H(H') \rightleftharpoons CH_3OH' + H^{+}$$

(4) Hydrogen Bonding

Unlike most chemical shifts for CH protons, the chemical shifts for OH and NH protons depend on the physical environment. For example, these chemical shifts are temperature-dependent. Of more practical importance to most organic chemists is that the chemical shifts of OH and NH protons are *concentration-dependent* because of hydrogen bonding. In a nonhydrogen-bonding solvent (such as CCl_4) and at low concentrations (1% or less), OH proton absorption is observed at a chemical shift of around 0.5 ppm. At the more usual, higher concentrations, the absorption is observed in the 4–5.5 ppm region. The OH proton absorption can be shifted even farther downfield by hydrogen-bonding solvents.

16.3 Use of Deuterium in Proton NMR Spectroscopy

Deuterium nuclei do not absorb energy of the same frequency range as do protons; therefore, they do not show up in a proton NMR spectrum. This has several important consequences. NMR solvents are often deuterated so that they do not interfere with proton NMR spectra. Chloroform-d (deuteriochloroform, $CDCl_3$) is a common solvent for NMR work.

Another use of deuterium in NMR spectra is the substitution of a deuterium atom for a hydrogen atom in a compound to simplify an otherwise complex spectrum or to identify a particular proton. For example, the substitution of deuterium can be used to identify a peak arising from OH or NH protons. The sample is shaken with a few drops of D_2O, the NH or OH protons undergo chemical exchange, and the peak in the NMR spectrum that changes is thus identified as the OH or NH peak. In this case, a new peak will appear at a chemical shift of about 5 ppm because of the formation of DOH.

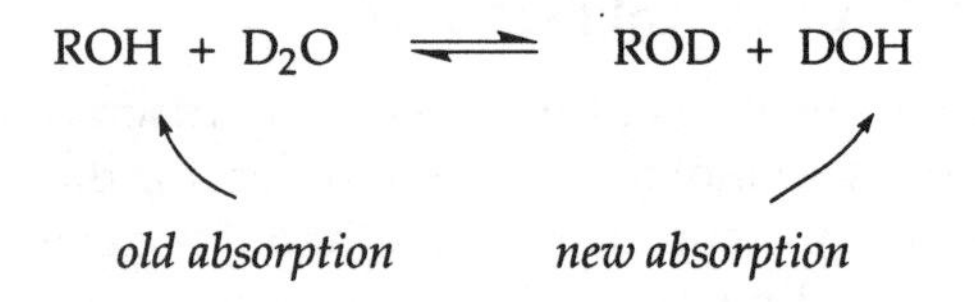

16.4 NMR Spectra of Unknowns

If a NMR spectrum shows well-spaced multiplets, you may be able to arrive at the structure of an unknown compound from its spectrum and a minimum amount of other information. On the other hand, if the spectrum shows extensive fine structure, your chances of deducing the structure from the NMR spectrum alone are probably slim.

In a spectrum with well-spaced multiplets:

(1) Identify the different proton groups in the spectrum (alkyl, aryl, etc.), using their chemical shifts as a guide.

(2) Determine which protons are coupled. (Measure the J values, if necessary, to be sure they match.)

(3) From the relative areas under the peaks (the integration values), calculate the relative ratios of the different types of protons.

(4) Draw up a list of partial structures consistent with each of the singlets or multiplets, their chemical shifts, and their relative areas. In drawing up this list, consider *all* the information available—infrared spectrum, elemental analysis (Appendix II), etc. If you have the molecular formula, a calculation of the *degree of unsaturation* is helpful because it gives you the number of rings and/or pi bonds in the molecule (see Appendix II, p. 211).

(5) Consider the list of partial structures as a jigsaw puzzle. Fit the pieces together, trying various combinations. Your final solution must be consistent with *all* the data.

If the spectrum is too complex for complete interpretation, then simply glean as much information as you can from the spectrum. In general, you should attempt to identify the types of protons (alkyl, aryl, etc.). Calculate the ratios of the various groups of protons from the relative areas of the multiplets. Determine which types of protons are absent in the spectrum as well as those present. For example, if the spectrum shows no absorption for aryl protons, you can eliminate the phenyl group as part of the structure. Take what clues you can, and then turn your attention to other sources of information, such as the infrared spectrum, physical constants, and chemical information.

16.5 The NMR Spectrometer

Modern instruments are interfaced with a computer and employ Fourier Transform (FT) spectroscopy, or FT-NMR. These instruments excite all of the nuclear spin transitions simultaneously at a fixed applied field (B_0) by applying high-power radio frequency in a short pulse. This short-duration pulse contains a bandwidth of frequencies: all the frequencies necessary to excite all of the protons of an organic molecule simultaneously. The spectrometer monitors the magnetization from all of the excited nuclei as they decay back to the ground state. The resulting free-induction decay signal, or FID, is stored in digital form on the computer. Fourier Transform converts the FID to a frequency domain spectrum, which is the "normal" NMR spectrum.

Many pulses are made in a few minutes and the FIDs stored and averaged. If the background noise is large, the chemist can acquire a lot of scans to reduce the signal-to-noise ratio. FT-NMR instruments are fast, unlike the older continuous-wave instruments, and allow the chemist not only to acquire many scans for a single sample, but to run many different samples in a short amount of time.

Spectrometers employ a superconducting magnet called a solenoid to achieve very high applied magnetic fields. For organic chemistry applications, instruments capable of producing magnetic fields of 300–600 MHz are routinely used, while biochemists are investigating molecules using 700–800 MHz instruments. The sensitivity of the spectrometer is directly related to the magnetic field strength it produces—spectra acquired on the higher-MHz instruments tell the chemist more details about the molecule than those acquired on the low-field strength instruments. (Greater field strength also means the instrument costs more money.) Instruments that produce only 60-MHz fields are still in use, and many printed spectra in textbooks were generated on these older instruments.

Because an FID is in digital form, the data can be manipulated in many ways. Peaks can be expanded or enhanced to show details, values can be printed above the peaks, coupling constants can be measured, the area under the peaks can be integrated, and so on.

16.6 Preparing Samples for NMR Spectra

In some laboratories, students are expected to operate the NMR spectrometer; in other laboratories, students submit prepared samples to a spectrometer operator. Your instructor will specify the procedure used in your laboratory and will instruct you in the use of the instrument if you are to obtain your own spectra. At some institutions, a NMR operator will acquire the spectra, but students will sit at a computer terminal to "work up" and print the spectra themselves.

To prepare a sample for NMR spectroscopy, dissolve 10–50 mg of the sample in about 0.9 mL of a deuterated solvent, usually $CDCl_3$. FT-NMR instruments require that samples be run in a solvent containing deuterium because the instrument locks on the resonance of deuterium to achieve field–frequency stabilization. If the sample does not dissolve in $CDCl_3$, other deuterated solvents such as deuterioacetone (CD_3COCD_3), deuteriodimethylsulfoxide (DMSO-d_6), or deuteriobenzene (C_6D_6) may be used. Deuterated solvents are must be extremely pure and are expensive, with $CDCl_3$ being the least expensive as well as the most versatile.

The dissolved sample is then transferred to a special NMR tube. These tubes are manufactured to certain specifications of high-quality glass. In the instrument, the tubes are spun rapidly and thus must be balanced so that they spin evenly and do not break. When you prepare your sample, use only NMR tubes approved by the NMR operator.

All solid material must be removed from the solution before it is placed in the NMR tube. Suspended or solid particles cause broadening of the absorption peaks in the spectrum. Passing the solution through a Pasteur pipet that contains a small wad of cotton at the bottom removes any solid particles. After you prepare your sample, measure the depth of the liquid with a standard (properly filled) NMR tube and adjust the volume of your sample tube as necessary. For you to obtain a good NMR spectrum, it is important that NMR samples are filled to the proper depth.

A sample run in $CDCl_3$ will always show a very small peak at 7.24 ppm because deuteriochloroform is never 100% pure and a tiny amount of residual ordinary chloroform, $CHCl_3$, is in the sample. This peak is used to calibrate the spectrum during the workup process.

Additional Resources

A list of suggested readings which cover the theory of NMR spectroscopy is published on the Brooks/Cole Web site. Other useful resources on the web site include references to compilations of NMR spectra, links to tutorials, and practice problems. Please visit www.brookscole.com.

Problems

16.1 List the following protons in order of increasing chemical shift (smallest shift first).

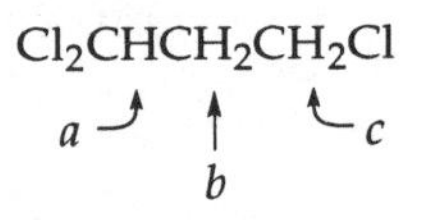

16.2 Pyrrole shows two principal CH signals in its NMR spectrum: at 6.5 ppm and 6.7 ppm. Is pyrrole an aromatic compound or a conjugated diene?

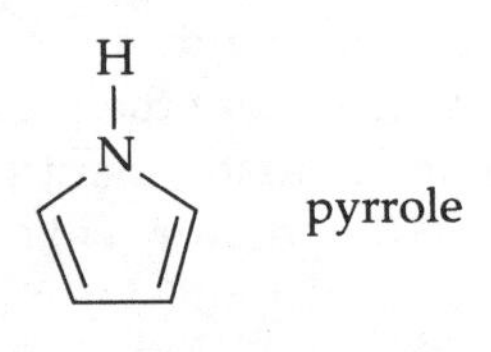

16.3 How many types of nonequivalent protons does each of the following compounds contain? (*Example:* CH_3CH_2Cl has two types, CH_3 and CH_2.)

(a) $(CH_3)_2CHCl$ (b) $CH_3CH(CH_2Cl)_2$

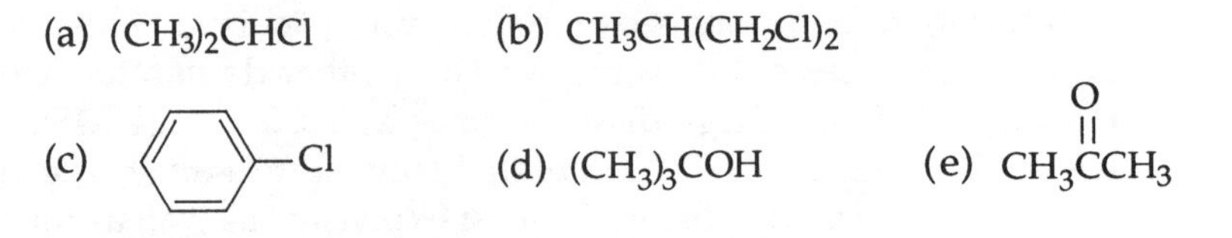

16.4 In the preceding problem, determine the relative areas under the principal signals for each compound.

16.5 What would be the splitting pattern (singlet, doublet, etc.) observed for each of the following indicated protons?

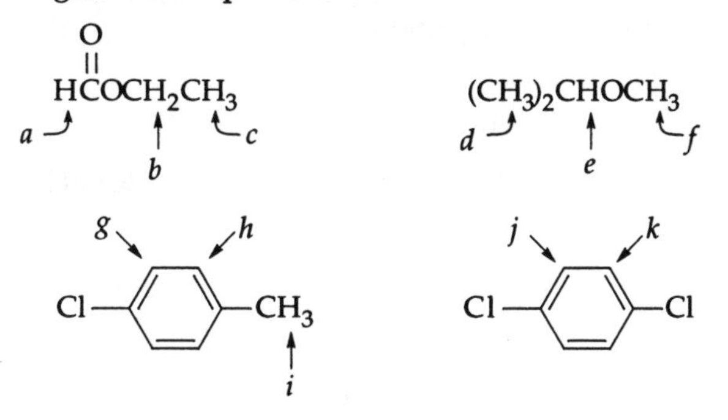

16.6 A NMR spectrum shows a singlet at 7.3 ppm, a triplet at 4.3, a triplet at 2.9, and a singlet at 2.0. The relative areas are 5, 2, 2, and 3, respectively. Which of the following compounds is compatible with this spectrum?

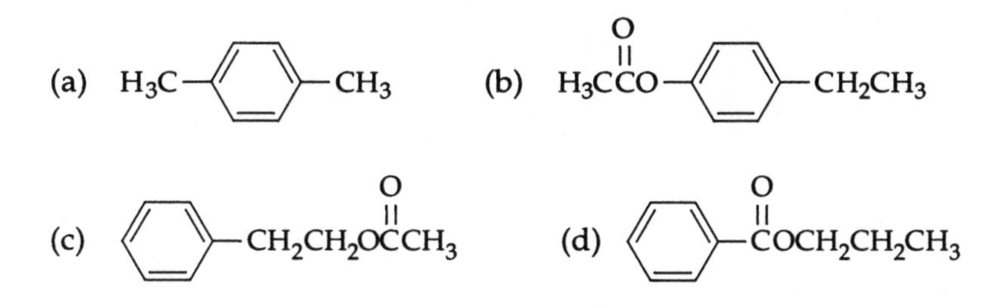

16.7 How would you distinguish between the following pairs of compounds using NMR spectroscopy? (Include in your answer the expected splitting patterns and relative areas of the NMR absorption.)

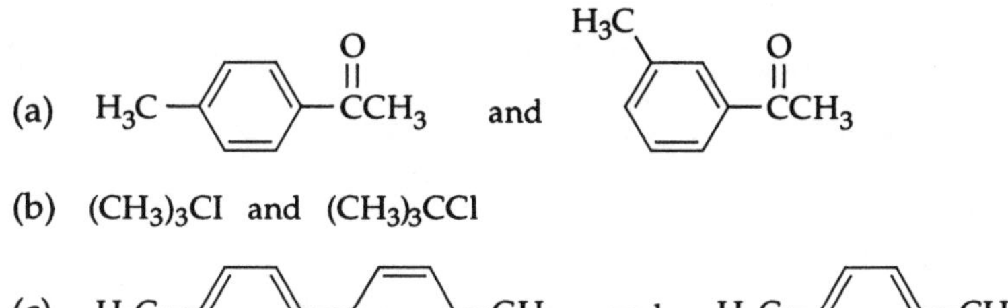

(b) $(CH_3)_3CI$ and $(CH_3)_3CCl$

(c) H_3C–C_6H_4–C_6H_4–CH_3 and H_3C–C_6H_4–CH_3

(d) $CH_3CO_2CH_2CH_3$ and $CH_3CH_2CO_2CH_3$

16.8 The following types of protons (underlined) all show NMR absorption between $\delta = 10\text{–}16$ ppm. How would you use the infrared spectrum to identify the type of compound?

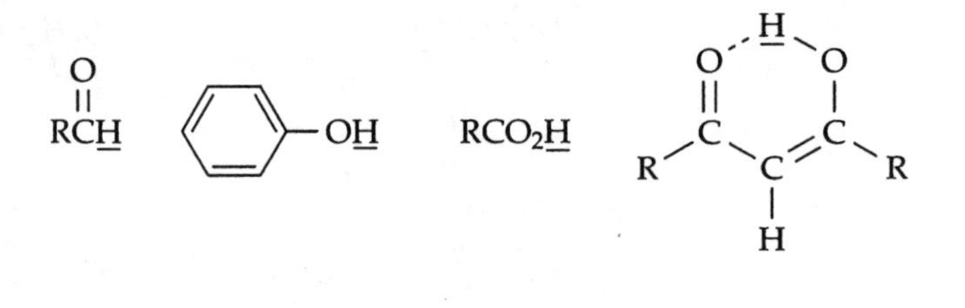

Technique 17

The Chemical Literature

17.1 Introduction to the Chemical Literature

The complete, written, published record of chemical knowledge is referred to as the **chemical literature**. The **primary literature**, or **original literature**, comprises the original reports of compound preparation, compound characterization, mechanistic studies, and so forth. These reports usually appear in research journals and in patent disclosures.

From the primary literature, information flows into the **secondary literature**. The secondary literature consists of

- compilations of data: either printed, bound handbooks, or online databases
- textbooks
- articles or books reviewing entire areas of research
- abstracts, or summaries, of individual research articles.

Figure 17.1 depicts the flow of information from the research laboratory through the literature.

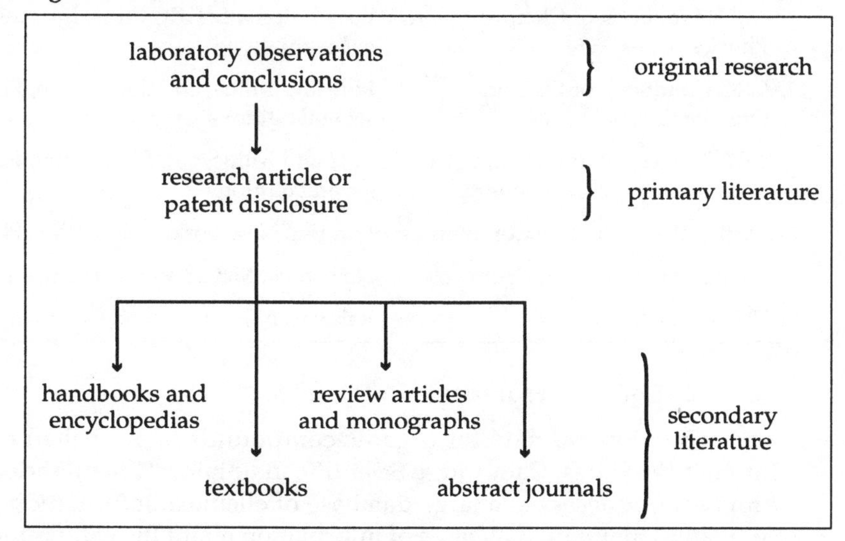

Figure 17.1 Flow of information into the chemical literature.

As a beginning organic chemistry laboratory student, your main interest in the chemical literature will be to find the physical constants of organic compounds, or to find information on reactions that you are studying in the laboratory or lecture course. (You will also be interested in finding hazard of organic compounds; this topic is covered in Appendix III, p. 213.) The sources for this level of information are in the secondary literature, especially the reference handbooks, online resources, and textbooks. If you are majoring in chemistry, you might want to read in-depth information as found in other secondary literature sources, such as review articles, monographs, and abstract journals.

17.2 Secondary Literature

A. Handbooks and Online Resources

(1) Handbooks

The physical constants of compounds are compiled in printed reference handbooks. These handbooks are usually found in the reference area of the science section of an academic library. Table 17.1 lists several of these handbooks. Be aware that the name of a compound as listed in an experimental procedure may not correlate with the nomenclature system used in a particular handbook. To find the compound, try using the molecular formula index provided in the handbook.

Table 17.1 Reference handbooks that contain physical constants.

Title	*Publishing information*
Aldrich Catalog/Handbook of Fine Chemicals	Published annually by the Aldrich Chemical Co., Milwaukee, WI (Phone 800-558-9160 for a copy.)
CRC Handbook of Chemistry and Physics	Boca Raton, FL: CRC Press, Inc., published annually or biennially
CRC Handbook of Data on Organic Compounds	Lide and Milne, ed., Boca Raton, FL: CRC Press, Inc., several editions
CRC Names, Synthesis, and Structures of Organic Compounds	Lide and Milne, ed., Boca Raton, FL: CRC Press, Inc., several editions
Dictionary of Organic Compounds	London; New York: Chapman & Hall, many editions
Lange's Handbook of Chemistry	New York: McGraw-Hill, many editions
The Merck Index	Rahway, N.J.: Merck and Co., many editions

(2) Online Resources

The physical data for organic compounds is published on many different Internet Web sites. Cambridge Scientific maintains "ChemFinder," a Web site that provides free access to a large database of chemical information. Chemical supply companies maintain databases of information about the compounds they sell. Acros Chemicals, Sigma-Aldrich, Baker Chemicals, and Mallinckrodt are examples of companies that offer free access to their chemical databases on the Internet. Many companies and agencies currently publish MSDS's (Material Safety Data Sheets; see

p. 214) on the Internet. MSDS's include the physical data as well as the hazard information for a compound.

Online versions of databases such as *The Merck Index* are available at many college or university libraries. Access to these databases is paid for by the academic institution; therefore, you must access them from the university's computer terminals. Check with the science librarian to find out what types of services are available at your institution.

Chem Abstracts Registry Numbers. Searching for a compound either online or in printed handbooks is sometimes easier if you use the Chem Abstracts Registry Number, or CAS RN. This is an identification code assigned by Chemical Abstracts Service (CAS) to a unique substance when it is added to the CAS registry system. You can think of a registry number as a "social security number" for a chemical. Although each compound only has one CAS RN, it can have many names. The molecular formula for a compound is not as useful in its identification, since there are many substances with the same molecular formula. For instance, there are 171 substances with the formula C_4H_8 (!), but only one registry number exists for each substance. Chemists have come to depend on the CAS registry number system for the absolute name, or identification, of a compound. The registry number has three parts, like this:

NNNNN-NN-N

The first series of "N" can be two to six digits in length, while the second series is always two digits and the last is always just one. For example, the CAS registry numbers for cyclohexanol and 2-propanol are

cyclohexanol 108-93-0
2-propanol 67-63-0

B. Textbooks

The best way to find information about a general organic chemistry topic is to start with a fairly general reference book such as the standard introductory organic chemistry textbooks. Currently, there are about 15 such texts in use in universities across the United States alone. If the one your instructor has assigned to you does not answer all of your questions, you should look for texts by different authors at the library, or ask your professor, laboratory instructor, or the storeroom personnel if organic chemistry texts are available for loan.

If you are interested in an *advanced* organic chemistry textbook, check with your professor for recommendations.

C. Review Articles and Monographs

Review articles are written by chemists who survey the current chemical literature for a single, limited topic and then condense and perhaps interpret the information for the readers. Several periodicals specialize in publishing review articles (see Table 17.2). Open-ended serial publications that contain review-type articles are published at irregular intervals in hardbound form. For example, *Organic Reactions* has detailed discussions of important reaction types (such as Friedel–Crafts alkylations) from the laboratory point of view.

Table 17.2 Some sources of review articles.

Periodicals specializing in review articles	*Open-ended serial publications*
Chemical Reviews	Progress in Physical Organic Chemistry
Chemical Society Reviews	Organic Reactions
Synthesis	Organic Reaction Mechanisms
Organometallic Chemistry Reviews	Topics in Stereochemistry
Accounts of Chemical Research	Advances in Organometallic Chemistry

Several books or series of books are devoted to synthetic methods or reagents used in organic reactions. Examples are *Organic Syntheses* (John Wiley, NY) and *Reagents for Organic Synthesis* (Fieser and Fieser, Wiley, NY). Monographs on specific organic chemistry topics are too numerous to list here. Check the library catalog at your university if you are interested in a particular topic.

D. Abstract Journals: Chemical Abstracts

In order to find an article on a particular chemistry topic, researchers first consult the secondary publications called **abstract journals**. These are periodicals that publish short abstracts of articles that have appeared in the original research journals.

The most comprehensive collection of modern chemical information is *Chemical Abstracts* (*Chem. Abstr., CA*). *Chemical Abstracts* has been published continuously since 1907. Today, the Chemical Abstracts Service abstracts hundreds of thousands of documents per year from about 14,000 periodicals. The collective index just for the five-year period 1977–1981 occupies over 131,455 pages! *Chemical Abstracts* covers all areas of chemistry and attempts to publish concise abstracts of every article and patent as soon as possible (usually three to four months from the original publication date). These abstracts are organized within each printed volume by general fields (organic chemistry, biochemistry, and so on).

Chemical Abstracts is available as printed books; however, it is also available at most universities as an online, searchable database. The service is called *SciFinder* and it is quite intuitive to use. It is probably accessible only through the computers in your academic library because it costs (the library) to use the service, as opposed to the free online databases listed above in Section A.

Beilstein's *Handbuch der Organischen Chemie* (Berlin: Springer-Verlag), commonly referred to as simply "Beilstein," is a monumental compilation of organic chemistry data. Beilstein was first published by Friedrich Karl Beilstein in 1881–1882. After the third edition, the German Chemical Society acquired the rights to this work and published the fourth edition, which is the edition used today. Beilstein is an excellent source of physical data for organic compounds as well as journal references to primary literature on these compounds. If you are interested in the printed Beilstein compilations, ask your instructor or the librarian for assistance. Access to an online version of Beilstein called *CrossFire/Beilstein* is also available at many academic science libraries. *CrossFire/Beilstein* is not as intuitive to use as *SciFinder* so you may require some assistance the first time you access the database. The academic online version of Beilstein available is a fee-associated service and covers all the years that Beilstein has been published. A free-access version of Beilstein is available on the Internet (ChemWeb); however, this database goes back only to 1980.

A few other searchable databases of chemistry abstracts are now available on the Internet, such as Web of Science and ChemWeb. These databases are free to users, although the journals represented are somewhat limited.

17.3 Primary Literature

Research journals or periodicals are the *primary* literature of chemistry. If you are researching a particular organic chemistry topic, you would begin in the secondary literature to find references to an article in the primary literature. For instance, search results from a secondary source such as the *Dictionary of Organic Compounds*, a review article, *Chemical Abstracts*, or Beilstein would direct you to a journal article in a research journal. Table 17.3 lists a few of the chemical research periodicals. These journals are available in printed form in many university libraries. Some are also available as full-text online articles, accessible through the university library, which pays a subscriber's fee.

Table 17.3 Some chemical research periodicals.

Name	*Abbreviation*
Journal of the American Chemical Society	J. Am. Chem. Soc.
Journal of Organic Chemistry	J. Org. Chem.
Canadian Journal of Chemistry	Can. J. Chem.
Journal of the Chemical Society	J. Chem. Soc.
Journal of Organometallic Chemistry	J. Organometal. Chem.
Organometallics	Organometallics
Tetrahedron	Tetrahedron
Tetrahedron Letters	Tetrahedron Letters

Additional Resources

An extensive bibliography of printed resources pertinent to the literature of organic chemistry is published on the Brooks/Cole Web site. The site also maintains links to Internet resources for physical data of organic compounds, educational sites, and chemical journal or journal abstract sites. Additional problems on the web site help guide students to information on chemicals that can be found on the Internet. Please visit www.brookscole.com.

Problems

17.1 Using the *Handbook of Chemistry and Physics,* find the melting points of the following compounds.
(a) *trans*-1,2-Cyclohexanedicarboxylic acid
(b) *m*-Aminophenol
(c) 2,5-Dibromofuran

17.2 Using *Lange's Handbook of Chemistry,* find the refractive index of each of the following compounds.
(a) Benzyl alcohol
(b) Formamide
(c) 3-Pentenoic acid chloride

17.3 Using the *Merck Index,* find the following information.
(a) The medical use of α-methyl-*p*-tyrosine
(b) The structure and primary literature reference to tocol
(c) What the Wenker ring closure is

17.4 Use *Chemical Abstracts* to find the CAS RN for the following compounds.
(a) 1-Cyclohexyl-2-propyn-1-ol
(b) Marvinol 2002

17.5 Use *Chemical Abstracts* to find the name and molecular formula of the compound with the CAS RN [1226-05-7].

17.6 Use *Chemical Abstracts* to find all articles that Tarek Sammakia published in 1996.

(Note: The next two problems require access to online resources.)

17.7 Write down the name of an ingredient listed on a bottle at home. The bottle can be a cleaning item, a medication, a food: whatever you wish. Search for the compound in the *Merck Index* or in *SciFinder.* Print the information you find about the compound.

17.8 Use the *Merck Index* and *SciFinder* to find out about one of the following topics. Go first to the word in *italics* and look it up in the *Merck Index*. Print the record and highlight the CAS registry number. Use this number to help you search *SciFinder* for the desired information. Print the abstract. Circle the document type; if it is a journal, highlight the journal name.
(a) Uses for *ferrocene*-treated zeolite catalysts
(b) The role of *benzoic acid* in the protection of trembling aspens
(c) Find information on the effectiveness of an arginine derivative of *ibuprofen* in pain management

Appendix I

Commonly Used Calculations

This appendix is a brief summary of calculations commonly performed in the laboratory. If you do not understand a concept presented here, you should review that concept in any good general chemistry text.

In any mathematical calculation, carry along the units of the numbers. In this way, you can determine which units cancel (and which do not cancel) as a check on how you have set up your equation. Also, before mixing any reagents or reporting any percent yield, always look at your calculations and ask yourself, "Is this reasonable?"

1 Molarity

Molarity is defined as *the number of moles of solute in 1.00 liter of solution*. The following equations are used to determine molarity.

$$\text{molarity } (M) = \frac{\text{moles of solute}}{\text{liters of solution}}$$

$$= \frac{\text{g of solute/MW of solute}}{\text{liters of solution}}$$

$$= \frac{\text{g of solute}}{\text{(MW)(liters)}}$$

EXAMPLE 1 What is the molarity of an aqueous solution of 5.0 g of NaOH in 100 mL of solution?

$$M = \frac{\text{g}}{\text{(MW)(liters)}} = \frac{5.0\text{ g}}{(40.0\text{ g/mol})(0.100\text{ L})} = 1.25\text{ mol/L, or } 1.25M$$

EXAMPLE 2 What weight of solid NaOH would you need to prepare 50 mL of a 2.0M aqueous solution?

$$M = \frac{\text{g}}{\text{(MW)(liters)}}$$

$$\text{g} = (M)\text{(MW)(liters)}$$

$$= (2.0\text{ mol/L})(40.0\text{ g/mol})(0.050\text{ L})$$

$$= 4.0\text{ g}$$

2 Normality

The **normality** of a solution is the number of *equivalents of solute in 1.00 liter of solution*. In the organic laboratory, normality is generally encountered only with acids and bases (for example, 6*N* HCl, 6*N* H_2SO_4, or 6*N* NH_4OH). One equivalent of acid or base is the weight of the substance that contains 1.00 mole of H^+ or OH^-.

- For a monoprotic acid (such as HCl) or a base containing one OH^- per molecule (such as NaOH):

  equivalent weight = molecular weight

  Therefore,

  normality = molarity

  Examples:

  6*N* HCl = 6*M* HCl
  3*N* NaOH = 3*M* NaOH

- For a diprotic acid, such as H_2SO_4, or a base containing two OH^- per molecule, such as $Ca(OH)_2$:

  $$\text{equivalent weight} = \frac{1}{2}\ \text{molecular weight}$$

  Therefore,

  normality = 2 x molarity

  Examples:

  6*N* H_2SO_4 = 3*M* H_2SO_4
  2*N* $Ca(OH)_2$ = 1*M* $Ca(OH)_2$

EXAMPLE 3 Calculate the molarity of a 2.5*N* aqueous solution of H_3PO_4.

In this case, 1 mole of H_3PO_4 can theoretically supply 3 moles of H^+. The normality is three times the molarity.

$$N = 3M$$

$$M = \frac{N}{3} = \frac{2.5N}{3} = 0.83$$

EXAMPLE 4 What weight of $Ca(OH)_2$ would be necessary to prepare 100 mL of a 0.25*N* solution?

$$N = \frac{\text{no. of equivalents}}{\text{liters}}$$

$$\begin{aligned}\text{no. of equivalents} &= N \times \text{liters}\\ &= (0.25\ \text{equivalent/L})(0.100\ \text{L})\\ &= 0.025\ \text{equivalent}\end{aligned}$$

$$\text{equivalent weight} = \frac{1}{2}\ \text{MW}$$

because $Ca(OH)_2$ can supply two OH^- ions per $Ca(OH)_2$ molecule. Therefore,

$$0.025\ \text{equivalent} = \frac{1}{2}\ (0.025\ \text{mol})$$

And since the molecular (formula) weight of $Ca(OH)_2$ is 74.10 g/mol,

$$0.025 \text{ equivalent} = \frac{1}{2}(0.025 \text{ mol} \times 74.10 \text{ g/mol}) = 0.93 \text{ g}$$

As a check on N–M conversions, remember that for a given solution **N is always equal to or larger than M.**

3 Dilutions

In practice, we are often required to dilute a more concentrated acid (or base) to a less concentrated solution. Because the number of moles or equivalents of acid (or base) is not changed by dilution, the following simple equations allow us to calculate the amount of more concentrated solution needed.

$$M_1V_1 = M_2V_2 \quad \text{or} \quad N_1V_1 = N_2V_2$$

where M_1V_1 and N_1V_1 refer to the concentrated solution, and M_2V_2 and N_2V_2 refer to the dilute solution.

EXAMPLE 5 What volume of 12M HCl is needed to prepare 100 mL of 1.5M HCl?

$$M_1V_1 = M_2V_2$$

$$V_1 = \frac{M_2V_2}{M_1}$$

$$= \frac{(1.5M)\,(0.100 \text{ L})}{12M}$$

$$= 0.0125 \text{ L, or } 12.5 \text{ mL}$$

4 Percent Concentrations

Many common laboratory manipulations require solutions with concentrations reported in percents. These percentages generally refer to **weight–volume percents.** For example, a 5% $NaHCO_3$ solution is an aqueous solution of 5 g of $NaHCO_3$ dissolved in water and then diluted *to* 100 mL (not *with* 100 mL).

$$\text{percent (weight/volume)} = \frac{\text{g of solute}}{100 \text{ mL of solution}}$$

EXAMPLE 4 What weight of NaOH is required to prepare 30 mL of a 15% aqueous solution?

This type of problem may be solved quickly by a simple proportion.

$$\frac{15 \text{ g}}{100 \text{ mL}} = \frac{x \text{ g}}{30 \text{ mL}}$$

$$x = \frac{15 \text{ g}}{100 \text{ mL}} \times 30 \text{ mL} = 4.5 \text{ g}$$

In certain instances, it may be desirable to convert a concentrated solution to a more dilute solution.

EXAMPLE 5 What volume of 5.0% $NaHCO_3$ is needed to prepare 7.0 mL of 2.0% $NaHCO_3$?

$$C_1V_1 = C_2V_2$$

$$V_1 = \frac{C_2V_2}{C_1}$$

$C_1 = 5.0\%$	$C_2 = 2.0\%$
$V_1 = ?$	$V_2 = 7.0$ mL

$$V_1 = \frac{2.0\% \times 7.0\text{ mL}}{5.0\%}$$

$$= 2.8\text{ mL}$$

5 Percent Yields and Theoretical Yields

Percent yield. Percent yield is simply the percent of the theoretical amount of product actually obtained in a reaction.

$$\text{percent yield} = \frac{\text{actual yield}}{\text{theoretical yield}} \times 100$$

EXAMPLE 6 What is the percent yield when 5.2 g of product are obtained from a theoretical 7.5 g?

$$\text{percent yield} = \frac{5.2}{7.5} \times 100$$

Theoretical yield. To calculate the theoretical yield, balance the reaction and calculate the moles of reactants. Then calculate the theoretical yield based on the limiting reagent, which is the reagent present in the shortest supply. For example, in the following oxidation of cyclohexanol, 20.0 g of the alcohol are treated with 23.8 g of $Na_2Cr_2O_7 \bullet 2H_2O$ and 26.5 g of H_2SO_4.

(1) Calculate or look up the molecular weights.

(2) Calculate the numbers of moles.

(3) Determine the limiting reagent. The required amounts of $Na_2Cr_2O_7 \bullet 2H_2O$ and H_2SO_4 to oxidize 0.20 mole of cyclohexanol are 0.20/3, or 0.067, mole of the chromate and (4/3)(0.20), or 0.27, mole of H_2SO_4. In this particular reaction, the alcohol and H_2SO_4 are limiting reagents, while the chromate is present in excess.

(4) Determine the theoretical number of moles of product possible. In this case, 0.20 mole of cyclohexanone is the maximum, or theoretical, yield (the same as the number of moles of cyclohexanol, one of the limiting reagents).

(5) Convert the theoretical yield of product to grams:

0.20 mol of cyclohexanol = 0.20 mol x 98.15 g/mol

= 19.6 g (theoretical yield)

reactants:

$$3\ \text{C}_6\text{H}_{11}\text{–OH} + NaCr_2O_7 \bullet 2H_2O + 4H_2SO_4 \longrightarrow$$

MW:	100.16	298.00	98.08
weight:	20.0 g	23.8 g	26.5
moles:	0.20	0.080	0.27

products:

$$3\ \text{C}_6\text{H}_{10}\text{=O} + Cr_2(SO_4)_3 + Na_2SO_4 + 9H_2O$$

MW:	98.15	—	—	—
weight:	?	—	—	—
moles:	?	—	—	—

Appendix II

Elemental Analyses

The **weight percents of carbon and hydrogen** in an organic compound are determined by burning a weighed sample of the compound in a special apparatus. The resultant water vapor is collected in a tared chamber containing a drying agent, while the carbon dioxide is trapped in a chamber containing a strong base (which converts the carbon dioxide to the carbonate ion). The chambers are reweighed and the weights of water and carbon dioxide determined by difference. The percent C and percent H in the original compound can then be calculated.

In practice, the problems of ensuring complete combustion and collecting 100% of the gases (uncontaminated by outside moisture or carbon dioxide) are difficult to overcome without the proper equipment. Therefore, most organic chemists do not perform these analyses but send samples to analytical chemists who specialize in this type of analysis. It is, of course, the responsibility of the organic chemist to submit a sample that is as pure as possible. Impurities amounting to 5% of the sample may not measurably affect an infrared or NMR spectrum, but they will invalidate the results of a C and H analysis.

Most organic compounds contain only C, H, and O. A typical analysis report shows only the percent C and percent H. The percent O is usually determined by difference.

If requested to do so, an analytical laboratory can perform analyses for oxygen, the halogens, phosphorus, nitrogen, and other elements. Here, we will discuss compounds containing only C, H, and O.

1 Determining the Empirical Formula

From the weight percents of the elements in a compound, the **molar ratio** of the elements can be calculated. This is accomplished by dividing each *weight percent* by the *atomic mass* of that element. The following example describes this procedure for a compound containing 38.72% C, 9.72% H, and 51.56% O.

Step 1 Divide percent by atomic mass to determine the molar ratio:

$$\frac{38.72\%\ \text{C}}{12.01} = 3.22 \qquad \frac{9.72\%\ \text{H}}{1.008} = 9.64 \qquad \frac{51.56\%\ \text{C}}{16.00} = 3.22$$

We see that the molar ratio of C, H, and O is 3.22:9.64:3.22. These numbers must be converted to small whole numbers by dividing all three values by the smallest value.

Step 2 Divide the values in the ratio by the smallest value:

$$\text{for C: } \frac{3.22}{3.22} = 1.00 \qquad \text{for H: } \frac{9.64}{3.22} = 2.99 \qquad \text{for O: } \frac{3.22}{3.22} = 1.00$$

When the molar ratio has been converted to small whole numbers, the ratio tells us the relative numbers of the atoms in a molecule. In our example, the molar ratio is very close to 1:3:1. From these numbers, we can write an **empirical formula**.

Step 3 Write the empirical formula:

CH_3O

Note that this formula is an empirical formula, which shows only the ratios of the atoms, not their actual numbers in the molecule. The true **molecular formula** might be CH_3O, $C_2H_6O_2$, $C_3H_9O_3$, or any other formula with C, H, and O in a ratio of 1:3:1.

EXAMPLE 1 A compound contains 65.50% C, 9.46% H, and 25.02% O. What is its empirical formula?

Step 1 Divide percent by atomic mass:

$$\frac{65.50\%\ \text{C}}{12.01} = 5.45 \qquad \frac{9.46\%\ \text{H}}{1.008} = 9.38 \qquad \frac{25.02\%\ \text{C}}{16.00} = 1.56$$

Step 2 Divide by the smallest value:

$$\text{for C: } \frac{5.45}{1.56} = 3.49 \qquad \text{for H: } \frac{9.38}{1.56} = 6.01 \qquad \text{for O: } \frac{1.56}{1.56} = 1.00$$

In this example, the values for the molar ratio at this point are not all close to small whole numbers but can be rounded to 3.5:6:1. Multiplying this ratio by 2 does result in the necessary small whole numbers that are needed for an empirical formula—in this case, 7:12:2. The empirical formula is therefore $C_7H_{12}O_2$.

2 Determining the Molecular Formula

To convert an empirical formula to a molecular formula, we need to know the **molecular weight** of the compound. An analytical laboratory can determine the molecular weight along with the elemental analysis. Freezing-point depression is one way to accomplish this. In many cases, an organic chemist can determine the molecular weight of a compound from its mass spectrum.

In a molecular formula determination, the molecular weight of the empirical formula is simply compared with the experimentally determined molecular weight. If the two weights are the same, the empirical formula represents the molecular formula. If the values are *not* the same, the experimental molecular weight should be a simple multiple of the molecular weight of the empirical formula. This molecular formula is determined by multiplying the empirical formula by this number.

EXAMPLE 2 The theoretical molecular weight of CH_3O is 31.03. The actual molecular weight of the compound in question is found to be 62.11, which is very close to 2 x 31.03. Therefore, the molecular formula for the compound is $(CH_3O)_2$, or $C_2H_6O_2$.

3 Interpreting the Results of Elemental Analysis

In research, an elemental analysis is often used as evidence to substantiate the identity of a proposed structure. In a chemical journal, an analysis might be reported as: "Anal. Calcd for $C_2H_6O_2$: C, 38.70; H, 9.74. Found: C, 38.79; H, 9.67."

Generally, an analysis that results in values within 0.3% of the calculated value for each element is an acceptable piece of evidence for structure proof. An extremely pure sample will usually analyze closer to the calculated values. An analysis that does not fall within 0.3% suggests either an impure sample or possibly a different compound from the one expected.

If a compound contains a large number of carbon and hydrogen atoms (often the case with molecules of biological or medical importance), it is difficult to determine with any accuracy the exact molecular formula from analytical data. For example, *cholesterol*, $C_{27}H_{46}O$, contains 83.87% C and 11.99% H. Cholesterol contains one double bond. When this double bond is hydrogenated, *cholestanol*, $C_{27}H_{48}O$, is obtained. Cholestanol contains 83.43% C and 12.45% H. The percent compositions of these two compounds are fairly close, as you can see—much closer than the percent compositions of comparable smaller molecules would be. It would be difficult to differentiate cholesterol and cholestanol by elemental analysis alone.

4 Degree of Unsaturation

The true molecular formula of a compound gives important structural information about the compound. Obviously, it gives the types of atoms present in a molecule. Also, from the number of different types of atoms present, one can calculate the **degree of unsaturation**. The degree of unsaturation presents useful information on how the atoms are put together in the molecule. It limits the structural units or functional groups consistent with a particular molecular formula. It is especially useful in determining the structure of a compound by spectroscopic methods (Techniques 15 and 16).

The degree of unsaturation is the number of rings and/or π bonds in a molecule. Often denoted by the symbol Ω, it can be calculated readily from the molecular formula by applying the following rules:

Rule 1: Replace all halogens in the molecular formula by hydrogens.
Rule 2: Omit oxygens and sulfurs.
Rule 3: For each nitrogen, omit the nitrogen and one hydrogen.

Application of these rules reduces the molecular formula in question to the molecular formula of the hydrocarbon that has the same degree of unsaturation. The degree of unsaturation of a hydrocarbon is easily deduced if one remembers that a saturated hydrocarbon has the formula C_nH_{2n+2}. Thus for the formula C_nH_m,

$$\Omega = \frac{(2n+2)-m}{2} = n - \frac{m}{2} + 1$$

EXAMPLE 3 Determine the degree of unsaturation of a compound with the molecular formula C_8H_8NOBr.

Rule 1, replace halogens with hydrogens: C_8H_9NO

Rule 2, omit oxygens: C_8H_9N
Rule 3, omit the nitrogen and one hydrogen: C_8H_8

Therefore,

$$\Omega = 8 - \frac{8}{2} + 1 = 5$$

or, five rings and/or π bonds (or any combination of the two). Ω greater than or equal to 4 doesn't demand, but should suggest, the possibility of an aromatic (benzene) ring.

Examples of structures consistent with C_8H_8NOBr are

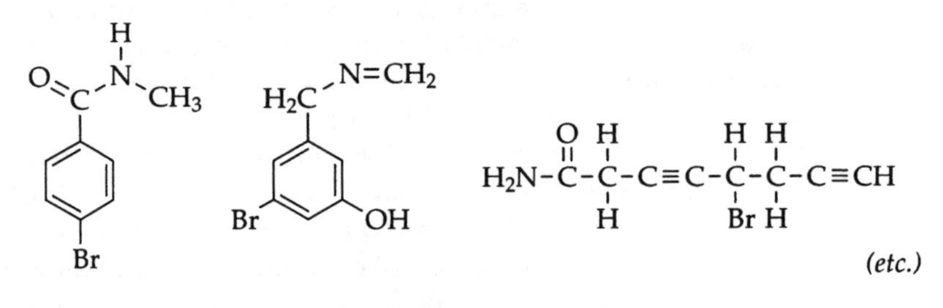

Appendix III

Health Hazards of Compounds Used in Organic Chemistry

Virtually every chemical has the potential for toxicity. Even dietary necessities can be poisonous when consumed to excess. Years ago, a Florida woman died from drinking too much water! Most compounds encountered in the organic laboratory are far more toxic than water. Yet, the toxicities of organic compounds began to be studied only in the second half of the 20th century, and only in the 1970s were comprehensive studies undertaken. Even today much of the available data is tentative or is applicable directly only to rats, mice, dogs, or monkeys, since it is obviously difficult to test toxicities with human subjects. Both government agencies and private corporations are still scrutinizing the allowable, "harmless" levels of organic compounds in the immediate environments of workers and consumers.

The federal government has developed a set of guidelines for workplaces (businesses or academic institutions) that use hazardous chemicals. These guidelines are meant to provide you, the worker or the consumer or the student, with all the hazard knowledge that is currently known about the chemicals, as well as training in their proper use.

Part of your organic chemistry learning experience is to gain a realistic knowledge of the hazards of chemicals. This understanding will help you in the student lab as well as a future workplace to minimize exposures as you handle chemicals. It will also help you in everyday situations, such as when you use pesticides, cleaners, paints, solvents, personal products, and even food additives.

1 Where to Find Chemical Hazard Information

A. Printed and Online Resources

Reference books like the *Merck Index* and the *Aldrich Catalogue/Handbook of Fine Chemicals* list the hazards of chemicals along with the physical properties. The Internet is a convenient place to find chemical hazard information. Several sites provided by government agencies specialize in hazard information. Company Web sites provide hazard information for each chemical they sell. Cambridge Scientific sponsors the ChemFinder Web site, which provides links to hazard information for a large database of chemicals.

B. Material Safety Data Sheets (MSDS)

Hazard information is provided in a chemical's Material Safety Data Sheet, or MSDS. According to OSHA (Occupational Safety and Health Administration) guidelines, an MSDS for every compound stored or used in the laboratory must be made available to you as a student in an academic institution. This is part of what is referred to as the **right to know** law. The official government document is titled Hazard Communication Standard (HCS) 29 CFR 1910.1200.* This standard stipulates that

- all workers (students) must know the hazards of chemicals used in the laboratories
- a list of all the chemicals present in the laboratories must be readily available
- an MSDS must be available for each chemical in the laboratory
- chemical containers must be properly labeled and their hazards must be listed
- all workers (students) must be trained in the proper handling of the chemicals

The items that must be included in an MSDS are listed in Table III.1. Since an MSDS is prepared by the manufacturer (or distributor), the format varies greatly depending on the vendor. All of the listed items *must* be included, but the order and organization are not specified. Thus, you will find that some MSDS's are easy to read, and that others are very difficult both to read and to understand.

Although your laboratory is required to have an MSDS for each compound it stores, it is not always practical for you to go to the laboratory to read it. Instead, you can find an MSDS for most compounds on Internet Web sites maintained by chemical companies (Acros, Baker, Sigma-Aldrich) or by universities (Cornell University, Vermont SIRI, Oklahoma State University).

C. Chemical Labels

The original vendor's labels on chemical containers are a good source of hazard information. The Hazard Communication Standard *requires* that original containers of chemicals as purchased from the manufacturer or distributor be labeled with a *precautionary label* that clearly states

- the identity of the chemical
- appropriate hazard warnings
- the name and address of the manufacturer or distributor

The vendor's label usually includes a lot more information than the three requirements listed above; for instance:

- impurities and other components
- flash point
- storage color code
- risk descriptive statement
- handling advice

---

* This is only a general overview of the law. Complete information about the OSHA requirements is available on the OSHA Web site (www.osha.gov).

Table III.1 Information that must be in an MSDS.

Required item	*Description*
Product identification	The chemical identity as listed on the label of the bottle, listing the percent composition of each ingredient. Hazardous ingredients must be noted. Common names must be listed for all hazardous ingredients.
Physical properties	The physical and chemical characteristics of the compound, including melting point, boiling point, flash point, etc.
Physical hazards	Physical hazards like fire, explosion, and reactivity data. Incompatibilities with other chemicals.
Health hazards	Health hazards, from both short- and long-term exposures. Toxicology data, such as LD, LC, TD, and TC values.
Exposure means	Primary route(s) of entry into the body.
Exposure limits	Exposure limits as set by OSHA or other agencies: PEL and TLV values.
Carcinogen information	Whether or not the chemical is a confirmed or potential carcinogen as determined by OSHA or other agencies.
Handling and storage	Precautions for safe handling and use.
Control measures	Applicable control measures, such as proper ventilation and personal protection.
First aid	Emergency and first-aid procedures in case of exposure.
Date	Date of preparation and latest revision of the MSDS.
Preparer	Contact information of the preparer of the MSDS.

- first-aid emergency medical advice
- NFPA hazard diamond
- recommended fire extinguisher class
- CAS number

Precautionary labels provide at a glance a good idea of the hazards associated with the chemical inside the container (see Figure III.1). The wealth of information provided on these labels is lost, however, when bulk chemicals are purchased and repackaged for internal use, such as for use in a classroom. At the very least, repackaged bottles should list the full chemical name of the compound and appropriate hazards.

On many bottles and in places of business, you may notice the diamond or the rectangle shown in Figure III.2. The National Fire Protection Agency (NFPA) diamond was developed so that firemen could readily ascertain the fire hazards of a chemical or of an area where chemicals are stored. The Hazardous Materials Identification Guide (HMIG) rectangle was developed by Lab Safety Supply and is similar to the Hazardous Materials Identification System (HMIS) used by the National Paint and Coating Association.

In each of the systems, red stands for fire, blue for health, and yellow for reactivity hazard. The relative hazard is designated as a number from 0–4, with 4 indi-

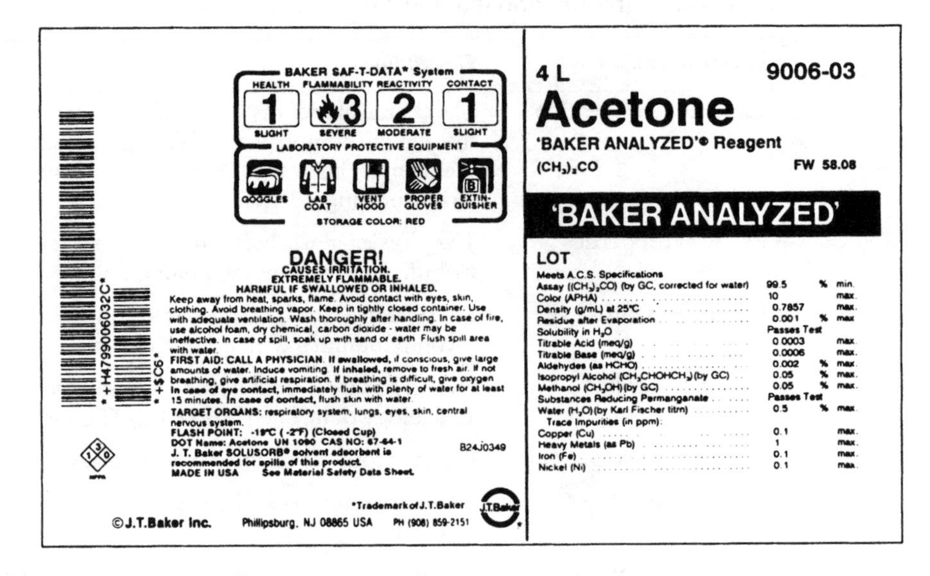

Figure III.1 A precautionary label. (Courtesy of J. T. Baker Chemical, Inc. Reprinted by permission.)

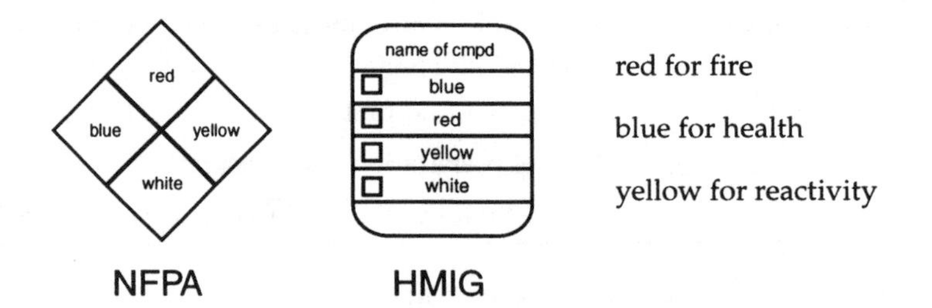

Figure III.2 NFPA and HMIG labeling symbols. Each colored area will contain a different number from 0–4, with 4 indicating the most severe hazard.

cating the most severe hazard. The white area is reserved in the NFPA system for special hazards. In the HMIG system, the white area directs the reader as to the personal protection equipment (PPE, see p. 3) recommended when handling the chemical.

2 Understanding Health Hazard Warnings

The health hazard warnings in books, in MSDSs, on Web sites, or on bottle labels will not help you unless you know their meanings. The following sections will explain the terms used to describe hazards of compounds. But first, let's examine the meaning of "toxicity."

There are several types of toxicity. A compound may be caustic and irritating, leading to burns and rashes. A compound may be relatively harmless in short-term exposure but may cause cancer if the exposure is repeated or prolonged. A short-term exposure to a compound may be harmless to an adult but cause serious defects in an unborn baby. (For this reason, pregnant women should pay strict attention to safety procedures, or even consider taking laboratory another semester.) Some compounds are deposited in the fatty tissue of the body instead of being eliminated and so may build up in concentration and lead to chronic toxicity. (This is frequently the case with organohalogen compounds.) The route of administration of a compound also affects its toxicity. Inhalation of vapors, aerosols, or dusts usually causes more severe symptoms than other types of exposure, such as ingestion or dermal absorption.

Toxicology studies themselves employ different techniques. One study may be concerned with the medical symptoms in a group of factory workers. Another study may involve feeding potentially toxic compounds to laboratory animals. Yet another piece of data may come from an isolated case reported in a medical journal.

Other problems that arise in toxicology are individual allergic reactions, variable individual tolerances, and differences in individual habits (for example, does the subject smoke or use any medication?). In time, comprehensive and directly comparable lists of toxicities may become available. Today, we must still extrapolate from one study to another, often with inadequate information.

Keeping all these variables in mind, let us consider some of the ways that toxicological data and health hazards are reported.

A. Toxicity

(1) LD and LC

LD stands for "lethal dose." Subscripts after the letters indicate the type of test. For instance LD_{50} means "lethal dose, 50% kill"—half of a statistical sample of test subjects (animals) died at a particular dose level. Another subscript designation is "Lo," for lowest toxic dose.

The route of administration used in the test will be designated as *inhalation, oral, dermal, iv* (intravenous), *im* (intramuscular), *ip* (intraperitoneal, or injected into the abdominal cavity), or *subcutaneous* (injected just under the surface of the skin).

All of the above information—route of administration, species tested, type of test, and toxicity levels—are necessary parts of any reliable report. The LD data for diethyl ether is reported as follows:

oral (rat) LD_{50}, 2200 mg/Kg

This means that when diethyl ether was administered through a tube into the stomachs of a statistical sample of rats at the amount of 2200 mg of ether per Kg of rat body weight, half the rats died. If only one rat from the statistical sample had died, the toxicity would be reported as the lowest lethal dose, or LD_{Lo} (oral, rats).

Animal lethal dose data can be roughly extrapolated to human toxicity. For instance, an animal LD_{50} of 50 mg/Kg correlates roughly to a 150-pound human ingesting about 1 teaspoon of a substance and the compound would be considered extremely toxic. An animal LD_{50} of 10 g/Kg correlates roughly to a 150-pound human ingesting about 3 cups of a substance and the compound would be considered only slightly toxic.

LC stands for "lethal concentration". It is similar to "lethal dose", but the "dose" is reported as a concentration. Concentration values are generally used for amounts of the compound in the air (see the discussion below).

(2) TD and TC

TD stands for "toxic dose," meaning an ill effect (as compared to LD, which is a "killing" dose). The same subscripts, routes of administration, and species tested delineations are used as with LD. For instance, if only one rat showed any ill effects at all when exposed to a substance at a certain amount, then the toxicity would be reported as the lowest toxic dose, or TD_{Lo} (oral, rats). TD_{50} means that 50% of the test subjects suffered from one or more toxic effects. TD data cannot be reported for humans (for obvious reasons).

TC stands for "toxic concentration." Toxic concentration values are a yardstick of the *inhalation* toxicity of a compound and therefore are of great importance to industrial workers and chemists exposed to fumes and dust. Toxic concentrations are generally reported as parts per million (ppm) or occasionally as parts per billion (ppb). An inhalation concentration of 1.0 ppm is 1.0 µL of sample per liter of air (1µL is 1.0×10^{-6} L).* The term TC_{Lo} means the lowest concentration known to cause any toxic effects in one or more members of the group of subjects.

B. Exposure Limits

Exposure limits are the amounts of a chemical to which people can be exposed without having adverse effects. These values are sometimes derived from TC_{Lo} values.

(1) TLV and PEL

TLV is *threshold limit value* and is used by the American Conference of Governmental Industrial Hygienists (ACGIH). PEL is the *permissible exposure limit* and is used by the Occupational Health and Safety Administration (OSHA). Values for TLV and PEL are usually equivalent. Each is the average concentration of a chemical in the air to which most people can be exposed and show no ill effect. A TLV-TWA is the amount to which a person can be repeatedly exposed for 8 hours a day, day after day, without adverse effect. A PEL or TLV value refers only to inhalation toxicity, not to skin or eye contact or to ingestion.

(2) TLV-STEL and TLV-C

STEL stands for "short term exposure limit." Thus, TLV-STEL is the amount of chemical in the air to which people can be exposed for a short period (no more than 15 minutes) without harm. Further, there must be no more than four such exposures in an 8-hour day, they must be at least 1 hour apart, and the total exposure must not exceed the TLV-TWA. TLV-C is the ceiling limit, or the maximum amount that can be in the air during any part of the working day.

---

* As a liquid concentration, the term "1.0 ppm" usually means 1.0 mg of solute in 1.0 L of solution.

Not all TLVs have been established by scientific experimentation. Many are estimates, based on experience with the chemical or based on known information about similar chemicals.

What do TLV values mean to students in a laboratory course? A TLV of 10 ppm or less indicates a particularly hazardous compound. If you use this compound in a certified fume hood, you can safely handle the compound. You should not use a large amount of a compound with a low TLV in the open air, especially if the room is not well ventilated. Most compounds used in student laboratories have TLVs of 20–100 ppm. See Table III.2 for the toxicities and exposure limits of some common solvents.

Table III.2 Toxicities of some common organic solvents.*

Name	*Toxicity (mg/kg)*	*TLV-TWA (ppm)*	*Other health hazards*
methylene chloride	oral rat LD_{50}: 1600	50	irritant; suspected carcinogen
methanol	oral rat LD_{50}: 5628	200	poison
acetone	oral rat LD_{50}: 5800	500	irritant
benzene	oral rat LD_{50}: 930	10	carcinogen
diethyl ether	oral rat LD_{50}: 1215	400	irritant
ethanol	oral rat LD_{50}: 7060	1000	
hexanes	oral rat LD_{50}: 28710	500	irritant
isopropyl alcohol	oral rat LD_{50}: 5045	400	irritant
ethyl acetate	oral rat LD_{50}: 5620	400	irritant
chloroform	oral rat LD_{50}: 908	10	irritant; suspected carcinogen

* Sources of data: MSDS's on the Mallinckrodt Web site (www.mallchem.com; accessed 4/2000), except for the data for benzene, which is from the Sigma-Aldrich Web site (www.sigma-aldrich.com; accessed 4/2000).

C. Contact Hazard

The contact hazard is usually expressed in terms of descriptive word form. *Corrosive* means that the chemical causes visible destruction of tissue if it comes into contact with skin. *Irritant* means that it causes a reversible inflammatory effect on living tissue. *Sensitizer* means that with repeated exposures it causes a substantial proportion of exposed people to develop an allergic reaction.

Often a contact hazard includes the affected body part, such as "highly irritating to the eyes" or "causes irritation of the respiratory tract."

D. Carcinogens

A carcinogen is a substance that can cause cancer. Officially, a compound must be labeled as a carcinogen if it is identified as such in the latest edition of one of three lists:

- National Toxicology Program (NTP) *Annual Report on Carcinogens*
- International Agency for Research on Cancer (IARC) Monographs
- OSHA 29 CFR1910, Subpart Z, Toxic and Hazardous Substances

On a more informal basis, compounds are called carcinogens if they are on the NTP or IARC lists of known *or* suspected carcinogens, on the "California List of Carcinogens," or on the list of the EPA Carcinogen Assessment Group.

Some known carcinogens include 4-aminobiphenyl, asbestos, benzene, chloromethyl methyl ether, cadmium, chromium[VI] compounds, Epstein-Barr virus, ethylene oxide, human immunodeficiency virus type 1 (infection with), human papillomavirus type 16, melphalan, radon, solar radiation, tamoxifen, and vinyl chloride.

E. Other Hazards

A *teratogen* is a compound that can cause fetal damage. The list of known or suspected teratogens is too long to include here; some of the more common ones are listed in Table III.3. The *allergens* in the same table include only a few of the compounds that frequently cause allergic reactions; unfortunately, a particular individual may be allergic to a compound that has no effect on most other people.

The fact that many toxic compounds can be absorbed *dermally*—through the skin—was largely ignored until recent years. For example, a lethal dose of phenol, once used as a surgical antiseptic, can be absorbed dermally. A laboratory worker would be wise to assume that all compounds can be absorbed dermally. In many cases, a relatively harmless solvent can carry other, nonabsorbable compounds through the skin. Dimethyl sulfoxide (DMSO), which a person can actually taste a after it has been applied to his or her hand, has been used to administer nonabsorbable drugs dermally. This solvent-carrying effect is a good reason not to use a solvent to cleanse your hands of other organic compounds unless absolutely necessary.

Additional Resources

The Brooks/Cole Web site maintains current links to Internet resources for hazards of organic compounds, MSDS's, explanations of chemical labels, lists of carcinogens, and other useful information. Please visit www.brookscole.com.

Table III.3 Examples of teratogens, allergens, and dermally absorbed compounds.

Known or suspected teratogens	*Allergens*	*Compounds that can be absorbed dermally*
benzene	pyrethrums	methanol
toluene	diazomethane	isopropyl alcohol
xylene	*p*-phenylenediamine	1-pentanol
aniline	some gums, glues, resins	2-chloroethanol
nitrobenzene		dimethyl sulfoxide
phenol		acrylonitrile
vinyl chloride		benzene
formaldehyde		nitrobenzene
dimethylformamide		bromobenzene
dimethylsulfoxide		phenol
N,N-dimethylacetamide		aniline (and some other aryl amines)
carbon disulfide		

Index